AF389433

Applied (Meta)-Heuristic in Intelligent Systems

Applied (Meta)-Heuristic in Intelligent Systems

Editor

Peng-Yeng Yin

MDPI • Basel • Beijing • Wuhan • Barcelona • Belgrade • Manchester • Tokyo • Cluj • Tianjin

Editor
Peng-Yeng Yin
Ming Chuan University
Taiwan

Editorial Office
MDPI
St. Alban-Anlage 66
4052 Basel, Switzerland

This is a reprint of articles from the Special Issue published online in the open access journal *Applied Sciences* (ISSN 2076-3417) (available at: https://www.mdpi.com/journal/applsci/special_issues/applied_heuristic_Intelligent_systems).

For citation purposes, cite each article independently as indicated on the article page online and as indicated below:

LastName, A.A.; LastName, B.B.; LastName, C.C. Article Title. *Journal Name* **Year**, *Volume Number*, Page Range.

ISBN 978-3-0365-5875-2 (Hbk)
ISBN 978-3-0365-5876-9 (PDF)

Contents

About the Editor

Peng-Yeng Yin

Peng-Yeng Yin is a Professor at the Information Technology and Management Program, Ming Chuan University, Taiwan. He has been a visiting Professor at the University of Maryland, Georgetown University, the University of California, Riverside (UCR), the University of Colorado, and Osaka University. He was a Distinguished Professor of the Department of Information Management, National Chi Nan University, Taiwan, in 2003 and 2021, and the Dean of the Information College at China University of Technology in 2022. Dr. Yin is the Editor-in-Chief of the *International Journal of Applied Metaheuristic Computing* and has been on the Editorial Board of the *Journal of Computer Information Systems*, *Applied Mathematics & Information Sciences*, *Mathematical Problems in Engineering*, *Mathematical Biosciences and Engineering*, *Applied Sciences*, *Symmetry*, *Sustainability*, and the *International Journal of Advanced Robotic Systems*. His research interests include artificial intelligence, metaheuristics, and operations research.

Editorial

Applying Modern Meta-Heuristics in Intelligent Systems

Peng-Yeng Yin

Information Technology and Management Program, Ming Chuan University, No. 5 De Ming Rd., Gui Shan District, Taoyuan City 333, Taiwan; pyyin@mail.mcu.edu.tw

Citation: Yin, P.-Y. Applying Modern Meta-Heuristics in Intelligent Systems. *Appl. Sci.* **2022**, *12*, 9746. https://doi.org/10.3390/app12199746

Received: 23 September 2022
Accepted: 26 September 2022
Published: 28 September 2022

Publisher's Note: MDPI stays neutral with regard to jurisdictional claims in published maps and institutional affiliations.

1. Introduction

Engineering and business problems are increasingly impenetrable due to the new economics triggered by big data, artificial intelligence, and the Internet of things. Exact algorithms and heuristics are not sufficient to conquer such extremely large and unstructured problems. Metaheuristic algorithms emerge as prevailing methods in this context. A generic metaheuristic framework guides the course of search trajectories beyond the local optimality, thus overcoming the impairment of traditional computation methods. The application of modern metaheuristics has a large coverage, ranging from unmanned aerial and ground surface vehicles, unmanned factories, resource-constrained production, and humanoids to green logistics, renewable energy, the circular economy, technology agriculture, environmental protection, finance technology, and the entertainment industry. The aim of this Special Issue is to collect new proposals for marrying modern metaheuristics and intelligent systems. The manuscript submission to this Special Issue was closed on 31 August 2022; we received 22 submissions, of which 8 papers were published, an acceptance rate of 36%. I believe the publication of this Special Issue will position itself at the research frontier across many aspects of applied sciences.

2. Modern Metaheuristics and Its Applications

In previous decades, we saw many novel metaheuristic algorithms including the genetic algorithm (GA), particle swarm optimization (PSO), ant colony optimization (ACO), differential evolution (DE), simulated annealing (SA), adaptive memory programming (AMP), and tabu search (TS). Recently, two interesting branches of metaheuristics have absorbed researchers' attention. Both of the two modern metaheuristics come from a marriage between two different disciplines, which can be easily recognized by their compound names: matheuristics and hyperheuristics. Matheuristics embeds mathematics into a metaheuristic framework, or vice versa. The purpose is to reach an elaborate mechanism melding together the effectiveness of mathematics and the efficiency of metaheuristics. The hyperheuristics aims to construct an automatic heuristic-selection machine, taking advantage of many lower-level heuristics which have been proposed in academia and industry.

This Special Issue presents two papers contemplating new matheuristics. The first paper, authored by P. Yin, P. Chen, Y. Wei, and R. Day, proposes a matheuristic approach embedding several memory programming strategies from AMP into the metaheuristic framework of the firefly algorithm [1]. The useful strategies include multiple guiding solutions, pattern search, multi-start, swarm rebuilding, and the objective landscape analysis. The second paper, authored by E. Cuevas, H. Becerra, H. Escobar, A. Luque-Chang, M. Pérez, H. Eid, and M. Jiménez, proposes matheuristic search schemes with the trajectory courses assisted by the second-order systems [2]. The second-order systems have different temporal responses depending on the set values of the parameters. Such temporally varying responses can be embedded into a metaheuristic to facilitate the search patterns adapted to complex landscapes. One hyperheuristics paper was chosen for this Special Issue. The paper, authored by X. Sánchez-Díaz, J. Ortiz-Bayliss, I. Amaya, J. Cruz-Duarte, S. Conant-Pablos, and H. Terashima-Marín, develops a feature-independent hyperheuristic approach for solving the knapsack problem. The proposed hyperheuristics does not rely on problem features to map the problem states into suitable existing heuristics [3]. Instead, a fixed sequence of existing heuristics is defined to improve the problem-solving performance within the hyperheuristic framework.

We also collected papers for classic applications of modern metaheuristics, namely vehicle routing and wireless networking. A paper authored by S. Nucamendi-Guillén, D. Flores-Díaz, E. Olivares-Benitez, and A. Mendoza presents a memetic algorithm for the cumulative capacitated vehicle routing problem [4]. The proposed method is a bi-objective optimization scheme that minimizes the total latency and total tardiness of the vehicle routing simultaneously. A mixed-integer program formulation is proposed for the cumulative capacitated vehicle routing problem. As compared to commercial software which optimally solves the problem with a small size, the proposed memetic algorithm can solve a larger-sized problem and produce the efficient bi-objective Pareto fronts. The paper by H. Rico-Garcia, J. Sanchez-Romero, A. Jimeno-Morenilla, and H. Migallon-Gomis proposes a parallel metaheuristic approach to reduce the vehicle traveling time in smart cities [5]. A Compute Unified Device Architecture (CUDA)-based implementation of the Teacher–Learner-Based Optimization (TLBO) metaheuristic is presented to target the shortest routing path for visiting a large number of points in a city. For applications in wireless networking, the paper by M. Chiang and W. Su presents a load balancing scheme for multithreaded applications on NUMA systems. When an imbalance occurs in the load on NUMA multiple cores, the load balancing mechanism of the kernel scheduler should migrate threads between NUMA cores. Threads to be migrated are selected considering the distribution of threads on nodes for inter-node load balancing [6]. The paper, authored by H. Zhang, J. Yang, T. Qin, Y. Fan, Z. Li, and W. Wei, develops a multi-strategy improved sparrow search algorithm (ISSA) for solving the node localization problem in heterogeneous wireless sensor networks (HWSNs). ISSA is an advanced version of the traditional SSA for improving the convergence speed and accuracy. This is accomplished by adopting the PSO's individual best for solution guiding and a Gaussian disturbance for preventing falling at local optima. The experimental results show that the ISSA yields a smaller average localization error than that of other metaheuristics [7].

Future challenges in metaheuristic applications are addressed in the final paper. The paper, authored by C. Huang, W. Liang, P. Chen, and Y. Chan, uses a dual matrix to identify the opinion leaders and followers in order to reach a converged consensus on the web [8]. The effectiveness of the proposed method has been validated by a case study of extracting leading opinions on green energy and low carbon for helping make effective public policies on environmental protection.

Funding: This research received no external funding.

Acknowledgments: I would like to thank all the contributing authors for their innovative and well managed works. This Special Issue would not be possible without their enthusiasm for charting advanced applications of metaheuristics in intelligent systems. My thanks are also given to the anonymous reviewers for their unselfish and professional reviews, which have helped shape the standard of this Special Issue.

Conflicts of Interest: The authors declare no conflict of interest.

References

1. Yin, P.; Chen, P.; Wei, Y.; Day, R. Cyber Firefly Algorithm Based on Adaptive Memory Programming for Global Optimization. *Appl. Sci.* **2020**, *10*, 8961. [CrossRef]
2. Cuevas, E.; Becerra, H.; Escobar, H.; Luque-Chang, A.; Pérez, M.; Eid, H.; Jiménez, M. Search Patterns Based on Trajectories Extracted from the Response of Second-Order Systems. *Appl. Sci.* **2021**, *11*, 3430. [CrossRef]
3. Sánchez-Díaz, X.; Ortiz-Bayliss, J.; Amaya, I.; Cruz-Duarte, J.; Conant-Pablos, S.; Terashima-Marín, H. A Feature-Independent Hyper-Heuristic Approach for Solving the Knapsack Problem. *Appl. Sci.* **2021**, *11*, 10209. [CrossRef]
4. Nucamendi-Guillén, S.; Flores-Díaz, D.; Olivares-Benitez, E.; Mendoza, A. A Memetic Algorithm for the Cumulative Capacitated Vehicle Routing Problem Including Priority Indexes. *Appl. Sci.* **2020**, *10*, 3943. [CrossRef]

5. Rico-Garcia, H.; Sanchez-Romero, J.; Jimeno-Morenilla, A.; Migallon-Gomis, H. A Parallel Meta-Heuristic Approach to Reduce Vehicle Travel Time in Smart Cities. *Appl. Sci.* **2021**, *11*, 818. [CrossRef]
6. Chiang, M.; Su, W. Thread-Aware Mechanism to Enhance Inter-Node Load Balancing for Multithreaded Applications on NUMA Systems. *Appl. Sci.* **2021**, *11*, 6486. [CrossRef]
7. Zhang, H.; Yang, J.; Qin, T.; Fan, Y.; Li, Z.; Wei, W. A Multi-Strategy Improved Sparrow Search Algorithm for Solving the Node Localization Problem in Heterogeneous Wireless Sensor Networks. *Appl. Sci.* **2022**, *12*, 5080. [CrossRef]
8. Huang, C.; Liang, W.; Chen, P.; Chan, Y. Identification of Opinion Leaders and Followers—A Case Study of Green Energy and Low Carbons. *Appl. Sci.* **2020**, *10*, 8416. [CrossRef]

Article

Cyber Firefly Algorithm Based on Adaptive Memory Programming for Global Optimization

Peng-Yeng Yin [1,2,*], Po-Yen Chen [1], Ying-Chieh Wei [2] and Rong-Fuh Day [1]

[1] Department of Information Management, National Chi Nan University, No. 1, University Rd., Puli, Nantou 54561, Taiwan; s100213516@ncnu.edu.tw (P.-Y.C.); rfday@ncnu.edu.tw (R.-F.D.)

[2] Institute of Strategy and Development of Emerging Industry, National Chi Nan University, No. 1, University Rd., Puli, Nantou 54561, Taiwan; s103245908@ncnu.edu.tw

* Correspondence: pyyin@ncnu.edu.tw

Received: 12 October 2020; Accepted: 14 December 2020; Published: 15 December 2020

Abstract: Recently, two evolutionary algorithms (EAs), the glowworm swarm optimization (GSO) and the firefly algorithm (FA), have been proposed. The two algorithms were inspired by the bioluminescence process that enables the light-mediated swarming behavior for mating or foraging. From our literature survey, we are convinced with much evidence that the EAs can be more effective if appropriate responsive strategies contained in the adaptive memory programming (AMP) domain are considered in the execution. This paper contemplates this line and proposes the Cyber Firefly Algorithm (CFA), which integrates key elements of the GSO and the FA and further proliferates the advantages by featuring the AMP-responsive strategies including multiple guiding solutions, pattern search, multi-start search, swarm rebuilding, and the objective landscape analysis. The robustness of the CFA has been compared against the GSO, FA, and several state-of-the-art metaheuristic methods. The experimental result based on intensive statistical analyses showed that the CFA performs better than the other algorithms for global optimization of benchmark functions.

Keywords: adaptive memory programming; firefly algorithm; global optimization; glowworm swarm optimization; metaheuristics

---

1. Introduction

Many challenging problems in modern engineering and business domains challenge the design of satisfactory algorithms. Traditionally, researchers resort to either mathematical programming approaches or heuristic algorithms. However, mathematical programming approaches are plagued by the curse of problem size and the heuristic algorithms have no guarantees to near-optimal solutions. Recently, metaheuristic approaches have come as an alternative between the two extreme approaches. The metaheuristic approaches can be classified into two classes, evolution-based and memory-based algorithms. The evolutionary algorithms (EAs) iteratively improve solution quality by decent operations inspired by nature metaphors, creating several novel algorithms, such as genetic algorithms, artificial immune systems, ant colony optimization, and particle swarm optimization. The memory-based metaheuristic approaches guide the search course to go beyond the local optimality by taking full advantage of adaptive memory manipulations. Typical renowned methods in this class include tabu search, path relinking, scatter search, variable neighborhood search, and greedy randomized adaptive search procedures (GRASP).

The metaheuristic approaches contained in the two classes have developed rather independently, and only a few works investigate the possible synergy between them [1]. Talbi and Bachelet [2] proposes the COSEARCH approach which combines the tabu search and a genetic algorithm for solving the quadratic assignment problem. Shen et al. [3] proposes an approach called HPSOTS which enables the particle swarm optimization to circumvent local optima by restraining the particle movement based on the use of tabu memory. A hybrid of the ant colony optimization and the GRASP is proposed by Marinakis et al. [4] for cluster analysis. However, the two metaheuristics are separately used to tackle the feature selection and clustering problems, respectively. Fuksz and Pop [5] proposes a hybrid genetic algorithm (GA) with variable neighborhood search (VNS) to the number partitioning problem. Their hybrid GA-VNS runs the GA as the main algorithm and the VNS procedure for improving individuals within the population. Yin et al. [6] introduces the cyber swarm algorithm which gives more substance to the particle swarm optimization (PSO) by incorporating the adaptive memory programming (AMP) strategies introduced in the scatter search and path-relinking (SS/PR) template. The adjective "cyber" emphasizes the connection between the evolutionary swarm metaheuristics and the AMP metaheuristics. The cyber swarm algorithm outperforms several state-of-the-art metaheuristics on complex benchmark functions. The experimental results of the above-noted works disclose a promising research area that the marrying of the approaches from the two classes of metaheuristics can create significant benefit that is hardly obtained by the approaches from each class alone.

More recently, two evolution-based algorithms [7,8], namely the glowworm swarm optimization (GSO) and the firefly algorithm (FA), were proposed. The two algorithms were inspired by the bioluminescence process that enables the light-mediated swarming behavior for mating or foraging. The intensity of the light and the distance between the light source and the observer determine the attractiveness degree that causes the moving maneuver. This form of metaphor can be used to develop a swarm-based optimization algorithm where a swarm of glowworms/fireflies represent a set of candidate solutions. The light intensity is evaluated by the objective value of the light source and the distance between the glowworms/fireflies implicitly defines the eligible neighbors since the light observed by an agent decays with the distance. The glowworms/fireflies are attracted by visible light sources and fly towards them. Therefore, the solutions represented by the glowworms/fireflies improve their objective value through the swarming behavior. Several improvements of FA have been proposed. Yang proposed the LFA [9] by combining his original FA with Levy flight. Yu et al. employed the variable step size strategy to create the VESSFA [10] variant. The dynamic adaptive inertia weight was adopted in WFA [11] to use the short-term memory of previous moving velocity. Kaveh et al. developed CLFA [12] which applies chaos theory and logical mapping to determine the optimal FA parameters. The Tidal Force formula was used in FAtidal [13] to improve the balance between exploitation and exploration search behaviors. The most recent improvement was GDAFA [14] which uses global-oriented positional update and dynamically adjusts the step size and attractiveness to avoid being trapped in local optima. The DsFA [15] employed dynamic step change strategy to balance the global and local search capabilities, such that the search course is not likely trapped in local optimum.

From a long-term perspective of metaheuristic development, we anticipate that the GSO and the FA can be made more effective by incorporating the notions from the memory-based approaches, as validated in many previous attempts. Based on the prevailing framework of the cyber swarm algorithm [6] that integrates key elements of the two types of metaheuristic methods, we propose the Cyber Firefly Algorithm (CFA) to proliferate the advantages of the original form of the GSO and the FA. The CFA conceptions include the employment of multiple guiding fireflies, the embedding of the pattern search, firefly swarm rebuilding by the multi-start search, and the responsive strategies based on objective landscape analysis. The robustness of the CFA has been compared against the GSO, FA, and several state-of-the-art metaheuristic methods. The result as demonstrated in our statistical analyses and comprehensive experiments showed the CFA manifests a more effective form of GSO and FA. Most FA variants intend to improve the position update mechanism such as by using Levy flight [9], variable step size strategy [10], adaptive inertia weight [11], and dynamically adjusting strategy [14]. Our CFA differs with these variants by facilitating a more intelligent step size control mechanism which performs landscape analysis in the objective space to adaptively tune the step size according to the profiles of the incumbent fitness landscape.

The novelty of this paper stems at creating a more effective form of GSO and FA approaches by bridging the advantages of evolution-based and memory-based metaheuristics. In particular, this paper investigates whether the CSA template is viable for improving GSO and FA as CSA has been shown in [6] for improving PSO. Our experimental results show that the proposed CFA prevails GSO, FA, and several state-of-the-art metaheuristics on benchmark datasets. This justifies the generalization capability of the CSA template and one can follow the template to create an effective version of interested metaheuristics.

The remainder of this paper is organized as follows. Section 2 presents a literature review and Section 3 proposes the CFA and the employed features. Section 4 presents the results of intensive experiments. Finally, conclusions and future research possibilities are given in Section 5.

2. Related Work Materials and Methods

2.1. Glowworm Swarm Optimization (GSO)

Krishnanand and Ghose [7] developed the GSO algorithm. It is assumed that each glowworm has a luciferin level and a local visibility range. The luciferin level of a glowworm determines its light intensity, and the local visibility range identifies the neighboring glowworms which are visible to it. The glowworm probabilistically chooses a neighbor which has a higher luciferin level than itself and is flying towards this neighbor due to light attraction. The local visibility range is dynamic for maintaining an ideal number of neighbors. The GSO simulates the glowworm behaviors and consists of three phases as depicted as follows.

The luciferin update phase evaluates the luciferin level of every glowworm according to the decay of its luminescence and the merit of its new position after performing the movement within the evolution cycle t. The luciferin level of glowworm i is updated by the following equation.

$$l_i^{t+1} \leftarrow (1 - \rho)l_i^t + \tau f_i^{t+1} \tag{1}$$

where ρ is the decay ratio of the glowworm's luminescence and τ is an enhancement constant. The first term is the persistence substance of luminescence due to decay with time, and the second term is the additive luminescence as a function of f_i^{t+1} which indicates the objective value measured at the glowworm's new position (here, without loss of generality, we assume the objective function is to be maximized).

In the movement phase of each evolution cycle, every glowworm in the swarm must perform a movement by flying towards a neighbor which has a higher luciferin level than the incumbent glowworm and is located within the local visibility neighborhood defined by a radius r_i^t. The probability with which glowworm i is attracted to a brighter glowworm j at evolution cycle t is given by:

$$p_{ij} = \frac{l_j^t - l_i^t}{\sum_{k \in N_i^t} l_k^t - l_i^t} \tag{2}$$

where N_i^t is the set of glowworms within the visibility neighborhood of glowworm i at evolution cycle t. Once selecting a neighbor, say glowworm j, the current glowworm i performs a movement to update its position as follows.

$$x_i^{t+1} \leftarrow x_i^t + s \left(\frac{x_j^t - x_i^t}{\|x_j^t - x_i^t\|} \right) \tag{3}$$

where s is the movement distance and $\|\bullet\|$ indicates the length of the referred vector. Precisely speaking, glowworm i moves in s units of distance towards glowworm j.

The visibility range update phase dynamically tunes the visibility radius of each glowworm to maintain an ideal number of neighbors, N^*. So, the current number of neighbors, $|N_i^t|$, is compared to N^* and the visibility radius r_i^t is tuned by the following equation.

$$r_i^{t+1} \leftarrow \min\left\{r_{\max}, \max\left\{0, r_i^t + \eta\left(N^* - N_i^t\right)\right\}\right\} \tag{4}$$

where $r_{\max}$ is the maximum visibility radius and η is a scaling parameter for tuning N_i^t. Therefore, the value of r_i^t is increased if $N_i^t < N^*$, and it is decreased if $N_i^t > N^*$. The feasible range of r_i^t is bounded between $[0, r_{\max}]$. The phenomenon of $r_i^t = 0$ indicates many glowworms are resorting to the position of the current glowworm, while $r_i^t = r_{max}$ discloses the situation that the current glowworm is in a large distance to most of the other glowworms.

2.2. Firefly Algorithm (FA)

Yang [8] introduced the FA. The FA explores the solution space with a population of fireflies. Each firefly has luminescence of flashing light which attracts its mates in an inverse multiplication of the squared distance and the light absorption. By using the metaphor, a firefly (representing a candidate solution) can improve its light intensity (a merit function of the objective value) by flying toward a more attractive firefly. In particular, the attractiveness of firefly j observed by firefly i is defined as follows.

$$\beta_r = \beta_0 e^{-\gamma r^2} \tag{5}$$

where γ is the light absorption coefficient, r is the Euclidean distance between the two fireflies, and β_0 is the light intensity of firefly j.

The movement of firefly i is attracted to a more attractive firefly j by the following equation,

$$x_i^{t+1} \leftarrow x_i^t + \beta_r\left(x_j^t - x_i^t\right) + \alpha\varepsilon_i \tag{6}$$

where the second term is due to the light attraction and the third term is a random perturbation with α being the randomization parameter and ε_i is a vector of small random numbers drawn from a Gaussian distribution or uniform distribution. After the movement, the light intensity of firefly i should be re-evaluated and the relative attractiveness of any other flies to firefly i is also re-calculated. The FA conducts a maximum number of evolution cycles and within each cycle every pair of fireflies should be examined for possible movement due to attraction and randomization.

2.3. Adaptive Memory Programming (AMP)

The AMP comprises a broader spectrum than its more popularly accepted branch, the tabu search [16]. In what follows, we focus our discussion on the use of the SS/PR template [17] which we found very effective in creating benefits for marrying with evolutionary-based metaheuristics.

2.3.1. Scatter Search (SS)

The scatter search (SS) [18] operates on a common reference set consisting of diverse and elite solutions observed throughout the evolution history. The SS method dynamically updates the reference set and systematically selects subsets of the reference set to generate new solutions. These new solutions are improved until local optima are obtained and the reference set is updated by comparing its current members to these local optima. Some important features of the SS are as follows. (1) A diversification-generation method is designed to identify the under-explored region in the solution space such that a set of diverse trial solutions can be produced. (2) An improvement method is used to enhance the quality of the solutions under keeping. This process usually involves a local search operation which brings the trial solution to a local optimum. (3) A reference set update method is adopted to make sure that the reference set is maintaining a set of high quality and mutually diverse solutions observed overall in search history. (4) The multi-start search strategy iteratively restarts a new search session when the current search loses it efficacy. To lead the search course towards under-explored solution space, the multi-start strategy works with a rebuilt reference set produced by the diversification-generation method. (5) Interactions between multiple reference solutions are systematically contemplated. The simplest implementation is to generate all 2-element subsets consisting of exactly two reference solutions. Various search courses are conducted between and beyond the selected reference solutions.

2.3.2. Path Relinking (PR)

There is a common hypothesis accepted by most of the metaheuristics that elite solutions often lie on trajectories from good solutions to other good solutions. In a broader sense, the crossover of chromosomes, the sociocognition learning of particles, and the pheromone trail searching are all effective operators following the noted hypothesis. PR therefore creates a search path between elite solutions. An initiating solution and a guiding solution are selected from the repository of elite solutions. PR then transforms the initiating solution into the guiding solution by generating a succession of moves that introduce attributes from the guiding solution into the initiating solution. The relinked path can go beyond the guiding solution to extend the search trajectory. PR works in the neighborhood space instead of the Euclidean space and variable neighborhoods are usually considered in performing successive moves. Therefore, PR is well fitted for use as a diversification strategy.

3. Proposed Methods

Our proposed CFA synergizes the strength from three domains, namely the GSO, the FA, and the AMP, to create a more effective global optimization metaheuristic algorithm. We articulate the new features of the CFA as follows.

3.1. Multiple Guiding Points

Both GSO and FA algorithms conduct the firefly movement by referring to a guiding point, which is a nearby elite firefly with a higher fitness than that of the moving firefly. This mechanism of using a single guiding point raises the risk of misleading due to the selection of a false peak. Our proposed CFA algorithm instead employs a two-guiding-point mechanism to enhance the exploration capability of the algorithm. More precisely, the neighboring fireflies of the incumbent firefly i are identified by reference to the current value of the visibility radius, r_i^t, where t indicates the index of the evolution iteration. Among the neighbors, the elite fireflies with a higher fitness than that of the incumbent firefly are eligible for guiding-point selection. We employ the rank selection strategy to select two guiding points from the eligible neighbors. According to our preliminary experiments, the rank selection strategy is more robust against prickly fitness landscape than the roulette-wheel selection strategy because the former works in the ranking value space instead of the fitness space.

3.2. Luciferin-Proportional Movement

In GSO and FA, the luciferin of firefly is used only for the determination of the single guiding point. As we deploy the two-guiding-point mechanism, the relative significance of contribution from each of the two guiding points can be further elaborated. We propose to convert the luciferin of each guiding point to the weighting value for individual contribution. Given two guiding points x_j^t and x_k^t observed at iteration t, their weighting values ω_1 and ω_2 are set as follows.

$$\omega_1 = \frac{l_j^t}{l_j^t + l_k^t} \tag{7}$$

$$\omega_2 = \frac{l_k^t}{l_j^t + l_k^t} = 1 - \omega_1 \tag{8}$$

We then propose the firefly movement formula as follows.

$$x_i^{t+1} \leftarrow x_i^t + \varphi_1 \omega_1 \beta_{r_j}\left(x_j^t - x_i^t\right) + \varphi_2 \omega_2 \beta_{r_k}\left(x_k^t - x_i^t\right) \tag{9}$$

where φ_1 and φ_2 are random values drawn from a uniform distribution U(*lb*, *ub*). Hence, firefly *i* is drawn by firefly *j* and firefly *k* with driving force relative to respective luciferin level. Parameters *lb* and *ub* control the feasible range of the step size for every firefly movement and their values are dynamically tuned in accordance with the landscape analysis as will be noted.

3.3. Adaptive Local Search

Local search is a rudimentary component contained in most of the successful metaheuristics such as memetic algorithms [19], scatter search [18], and GRASP [20], to name a few. Our CFA employs the pattern search method [21,22] which iteratively performs a two-step trajectory search. The explorative search step tentatively looks for an improving neighbor while the pattern search step aggressively expands the improving move by doubling the execution of the move. The proposed CFA embodies the pattern search method as a local search but performs it in an adaptive way. The local search is performed on all fireflies only after consuming every t_1 fitness function evaluations. To leverage the balance between search efficiency and effectiveness, we adaptively vary the frequency parameter (t_1) for the performed local search. The adaptive local search strategy is based on the analysis of the landscape observed in the objective space as will be noted.

3.4. Multi-Start with PR for Firefly Swarm Rebuilding

Multi-start is a mechanism for reinitiating the search with an under-explored region when the trajectory search loses its efficacy. The multi-start search strategy thus works with two functions. The function for *critical event detection* emits a signal upon the moment when the search stagnation behavior is observed. The *diversification-generation* function identifies an under-explored region in the solution space to generate a starting solution for the new trajectory search. Our CFA approach facilitates the critical event detection by monitoring the number of enduring iterations since the last improvement of the best-so-far firefly. If the number of no-improvement iterations for the best-so-far firefly has exceeded a parameter t_2, a signal for detection of a critical event is returned. The parameter t_2 is also made adaptive by the landscape analysis approach.

The diversification-generation function of the CFA approach is implemented by using the path-relinking technique, which has been found very useful in identifying a promising solution within an under-explored region and has been adopted as a diversification strategy such as in Yin et al. [6]. When a critical event signal is activated, a ratio δ of the swarm fireflies are rebuilt and each new firefly is placed on the best solution along a relinked path connecting a random solution and the best-so-far firefly. The path-relinking technique creates the path by dividing the subspace between the two end points of the path into *n* equal-size hyper-grids where *n* is the number of decision variables for the addressed optimization problem. A solution is sampled within each hyper-grid, so in total *n* solutions will be obtained along the relinked path. The best of the *n* sampled solutions is designated as the rebuilt firefly.

3.5. Responsive Strategies Based on Landscape Analysis

The previous features of CFA can be more effective by incorporating the notion of responsive strategy which adaptively tunes the search strategy when observing the status transitions. The CFA adapts the strategy parameters when the search encounters the transition between the smooth and prickly landscapes manifested by the objective function. The measure of fitness distance correlation (FDC) proposed by Jones and Forrest [23] has been proved useful for judging the suitability of a fitness landscape for various search algorithms. The FDC measures the correlation between the solution cost and the distance to the closest global optimum. Let the set of observed cost-distance pairs be $\{(c_1, d_1), \ldots, (c_m, d_m)\}$, the correlation coefficient is defined as

$$\text{FDC} = \frac{\frac{1}{m}\sum_{i=1}^{m}\left(c_i - \bar{c}\right)\left(d_i - \bar{d}\right)}{\sigma_c \sigma_d} \tag{10}$$

where $\bar{c}$ and $\bar{d}$ are the mean cost and the mean distance, and σ_c and σ_d. are the standard deviations of the costs and distances. Hence, a significant FDC value has the implication that on average, the better the solution quality the closer this solution is to an optimal solution. It can be contemplated that a single-modal objective function would impose a high FDC value in contrast to a multi-modal objective function whose FDC value is closer to zero because of the irrelevance between the fitness and the distance.

Most existing research adopted the FDC technique as an off-line analysis for realizing the characteristics of the fitness landscape, while our CFA exploits the fitness landscape analysis as an online responsive strategy. We periodically conduct the fitness landscape analysis for every $1000n$ fitness function evaluations, where n is the number of problem variables. All the visited solutions ever identified as local optima in the trajectory of a firefly within this time period are eligible for the FDC value computation because these local optima are representative solutions produced in the evolution. As the global optimum is unknown, the best solution obtained at the end of this period is considered to be the best-known solution. The FDC value is computed by using these local optima and the best solution identified in the current time period of previous $1000n$ fitness function evaluations. Consequently, our CFA performs the adaptive landscape analysis which can predict the objective landscape within the current region explored by the firefly swarm and activate appropriate responsive strategies to make the search more effective. If the derived FDC value for the current time period is greater than a threshold (FDC > h_1), the local objective landscape is considered to be asymptotical single-modal, and we perform two responsive strategies as follows: (1) increasing the distance, on average, of the firefly movement by $lb = lb/\lambda$ and $ub = ub/\lambda$ ($0 < \lambda < 1$); (2) increasing the elapsed iterations for checking the feasibility of executing the local search and multi-start search by $t_1 = t_1/\lambda$ and $t_2 = t_2/\lambda$. The implication of the two responsive strategies is that when the local objective landscape is single-modal, the search may be more effective if the firefly movement distance is greater and the frequency for conducting the local search and the multi-start search is lower. On contrary, If the absolute FDC value for the current time period is less than a threshold (|FDC| < h_2), the local objective landscape is considered to be asymptotical multi-modal, and we perform two responsive strategies as follows: (1) decreasing the distance, on average, of the firefly movement by $lb = lb \times \lambda$ and $ub = ub \times \lambda$; (2) decreasing the elapsed iterations for checking the feasibility of executing the local search and multi-start search by $t_1 = t_1 \times \lambda$ and $t_2 = t_2 \times \lambda$.

3.6. Pseudo-Code of CFA

The pseudo-code of our CFA is elaborated in Figure 1. The new features of the CFA are emphasized in boldface for comprehensive descriptions.

1   Initialize.
    1.1   Set the initial value of the algorithmic parameters.
    1.2   Generate the initial swarm of N_S fireflies at random.
    1.3   Evaluate the fitness of each firefly.

2   Repeat until *Max_FE* fitness function evaluations have been executed.

    2.1   For each firefly x_i^t, perform **rank selection** to select two local guides x_j^t and x_k^t from those fireflies which are brighter than x_i^t and within its visibility range.

        2.1.1   perform the **luciferin-proportional** and **multi-guider** firefly movement by

$$x_i^{t+1} \leftarrow x_i^t + \varphi_1 \omega_1 \beta_{r_j} \left(x_j^t - x_i^t\right) + \varphi_2 \omega_2 \beta_{r_k} \left(x_k^t - x_i^t\right)$$

        2.1.2   Evaluate the fitness of firefly x_i^{t+1}

        2.1.3   After every period of t_1 consecutive iterations, perform the **adaptive local search** on x_i^{t+1} until a local optimum is reached.

        2.1.4   Update the visibility radius by

$$r_i^{t+1} \leftarrow \min\left\{r_{max}, \; \max\left\{0, \; r_i^t + \eta\left(N^* - N_i^t\right)\right\}\right\}$$

    2.2   **Multi-Start**: If the best-so-far firefly has not improved for t_2 consecutive iterations, rebuild a ratio δ of the firefly swarm. Each rebuilt firefly is produced by

$$x_i^{t+1} \leftarrow path\_relinking\left(random\ firefly, best - so - far\ firefly\right)$$

    2.3   **Responsive Strategies**: After every period of $1000n$ iterations, conduct the fitness landscape analysis and adaptively tune the parameters, including average moving distance, local search frequency, and multi-start frequency.

Figure 1. Pseudo-code of the CFA algorithm.

4. Experimental Results and Discussions

We have conducted intensive experiments and statistical tests to compare the performance of the proposed CFA algorithm and its counterparts. The experimental results disclose several interesting outcomes in addition to establishing the effectiveness of the proposed method. The platform for conducting the experiments is a PC with an Intel Core i5CPU and 8.0 GB RAM. All programs are coded in C# language.

4.1. Benchmark Test Functions and Algorithm Parameter Settings

We have chosen two benchmark datasets. (1) The standard benchmark dataset contains 23 test functions that are widely used in the nonlinear global optimization literature [6,18,24,25]. The detailed function formulas can be found in the relevant literature and they have a wide variety of different landscapes and present a significant challenge to optimization methods. The number of variables, domain, and optimal value of these benchmark test functions are listed in Table 1. Performance evaluation on this dataset is reported in terms of the mean best function value obtained from 30 repetitive runs. For each run, the tested algorithm can perform 160,000 function evaluations (FE) for ensuring that the tested algorithm has very likely converged. (2) The IEEE CEC 2005 benchmark dataset which is designed for unconstrained real-parameter optimization. We selected the most challenging functions and compared our CFA with the best methods reported in [26] under the same evaluation criteria described in the original paper. The implementation of the benchmark functions from both datasets is available in public library [27] in some programming languages such as MATLAB and Java. As we used C# for all the experiments, we modified the public codes to C# and embedded them into our main program.

Table 1. The number of variables, domain, and optimal value of the benchmark functions.

No.	Function Name	Number of Variables (r)	Domain	Optimal Value
1	Easom (2)	2	$[-10, 10]$	-1.0
2	Shubert (2)	2	$[-10, 10]$	-186.7309
3	Rosenbrock (2)	2	$[-30, 30]$	0.0
4	Zakharov (2)	2	$[-5, 10]$	0.0
5	De Jong (3)	3	$[-5.12, 5.12]$	0.0
6	Shekel (4, 5)	4	$[0, 10]$	-10.1532
7	Shekel (4, 7)	4	$[0, 10]$	-10.4029
8	Shekel (4, 10)	4	$[0, 10]$	-10.5364
9	Sphere (10)	10	$[-100, 100]$	0.0
10	Rosenbrock (10)	10	$[-30, 30]$	0.0
11	Rastrigin (10)	10	$[-5.12, 5.12]$	0.0
12	Griewank (10)	10	$[-600, 600]$	0.0
13	Zakharov (10)	10	$[-5, 10]$	0.0
14	Sphere (20)	20	$[-100, 100]$	0.0
15	Rosenbrock (20)	20	$[-30, 30]$	0.0
16	Rastrigin (20)	20	$[-5.12, 5.12]$	0.0
17	Griewank (20)	20	$[-600, 600]$	0.0
18	Zakharov (20)	20	$[-5, 10]$	0.0
19	Sphere (30)	30	$[-100, 100]$	0.0
20	Rosenbrock (30)	30	$[-30, 30]$	0.0
21	Rastrigin (30)	30	$[-5.12, 5.12]$	0.0
22	Griewank (30)	30	$[-600, 600]$	0.0
23	Zakharov (30)	30	$[-5, 10]$	0.0

To obtain the parameter settings of the CFA, GSO, and FA, various values for their parameters were tested with the standard benchmark dataset and the values that resulted in the best mean function value were used as the parameter settings as tabulated in Table 2. The parameter settings used by other compared metaheuristics will be presented in Sections 4.3.2 and 4.3.3 where the comparative experiments are presented.

Table 2. Parameter settings of the CFA, GSO, and FA.

Notation	Parameter Description	Set Value
	Common parameters of CFA, GSO, and FA:	
Num_Run	Number of independent runs	30
Max_FE	Maximum number of executed FE for each run	160,000
N_S	Number of fireflies	60
N^*	Ideal number of local neighbors	10
L	Lower bound of decision variable	See Domain in Table 1
U	Upper bound of decision variable	See Domain in Table 1
r_{max}	Maximum visibility radius of fireflies	$(U - L) \times 0.05$
	Additional parameters used by CFA:	
lb	Initial minimum step size (to be adaptively tuned)	$(U - L) \times 10^{-6}$
ub	Initial maximum step size (to be adaptively tuned)	$(U - L) \times 10^{-2}$
t_1	Number of evolution iterations for activating adaptive local search (to be adaptively tuned)	20
t_2	Number of non-improving iterations for activating multi-start search (to be adaptively tuned)	50
δ	Ratio of swarm fireflies to be rebuilt	0.3
h_1	FDC single-modal threshold	0.5
h_2	FDC multi-modal threshold	0.4
λ	Scaling factor for performing responsive strategies	0.5

4.2. Analysis of CFA Strategies

To understand the influence on performance of using various CFA strategies, we conduct experiments on the benchmark functions with several CFA variants as in the following subsections.

4.2.1. Selection Strategy for Multiple Guiding Solutions

In contrast to GSO and FA, the CFA selects multiple guiding solutions for performing the firefly movement. The firefly can circumvent the false peaks by relaxing the limitation that constrains the firefly to move toward the single best solution within neighborhood. To investigate the influence on performance of using various selection strategies for guiding solutions, we compare the variants of the CFA that employs the roulette-wheel selection, the tournament selection, and the rank selection, respectively. The comparative performance of the CFA variants is listed in Table 3. It is seen that for the test functions with fewer than 10 variables and the relatively simple functions (Sphere and Zakharov), all of the CFA variants work well and the mean best value obtained is very near the optimal value. For the harder and larger functions (Rosenbrock, Rastrigin, and Griewank with 10 or more variables), the CFA with the rank selection strategy finds a more effective solution for most of the functions than the CFA with the other two selection strategies. The rank selection outperforms the other selection strategies in tackling harder problems because it eliminates the fitness-scaling problem by working in the rank-value space instead of in the function-value space such that the firefly will not be severely misled by false but higher peaks.

Table 3. The mean best function value and the standard deviation obtained by using various selection strategies.

Functions	Roulette-Wheel Selection		Tournament Selection		Rank Selection	
	Mean	Std	Mean	Std	Mean	Std
Easom (2)	−1.0	0.0	−1.0	0.0	−1.0	0.0
Shubert (2)	−186.726	0.011381	−186.723	0.010061	−186.723	0.016331
Rosenbrock (2)	0.0	0.0	0.0	0.0	0.0	0.0
Zakharov (2)	0.0	0.0	0.0	0.0	0.0	0.0
De Jong (3)	0.0	0.0	0.0	0.0	0.0	0.0
Shekel (4, 5)	−8.74128	2.436384	−8.64565	2.5462	−8.09613	3.003044
Shekel (4, 7)	−9.91291	1.345608	−10.1835	1.11003	−10.2611	0.440954
Shekel (4, 10)	−10.532	0.013263	−10.511	0.130888	−10.3554	0.966638
Sphere (10)	0.0	0.0	0.0	0.0	0.0	0.0
Rosenbrock (10)	0.000052	0.000031	0.000206	0.000982	0.002511	0.009392
Rastrigin (10)	0.851670	0.878212	0.847894	0.930433	0.646338	0.800856
Griewank (10)	0.023954	0.015559	0.018670	0.010681	0.016450	0.009914
Zakharov (10)	0.0	0.0	0.0	0.0	0.0	0.0
Sphere (20)	0.0	0.0	0.0	0.0	0.0	0.0
Rosenbrock (20)	0.800503	1.594902	0.953237	1.679062	0.267050	0.994099
Rastrigin (20)	3.052760	1.833682	2.607486	1.544556	2.466102	1.765681
Griewank (20)	0.013792	0.020267	0.011079	0.010911	0.007213	0.012479
Zakharov (20)	0.0	0.0	0.0	0.0	0.0	0.0
Sphere (30)	0.0	0.0	0.0	0.0	0.0	0.0
Rosenbrock (30)	5.924423	3.057484	2.962683	2.773249	0.801094	1.598004
Rastrigin (30)	4.204611	2.066075	4.973816	1.826780	4.680690	2.632701
Griewank (30)	0.009349	0.012921	0.008202	0.012808	0.003364	0.007581
Zakharov (30)	0.000005	0.000005	0.0	0.0	0.0	0.0

4.2.2. Adaptive Strategy for Local Search

Local search is a rudimentary component contained in most of the modern metaheuristic algorithms. The local search procedure exploits the regional function profiles and makes the master metaheuristic algorithm more effective than the original form which does not embed this procedure. However, the local search procedure can be computationally expensive if it is performed within each iteration of the master metaheuristic algorithm. As previously noted, the CFA applies the adaptive local search strategy which dynamically varies the frequency of the local search execution according to the result of the landscape analysis. Table 4 tabulates the mean best function value and the standard deviation obtained by CFA with or without adaptive local search. We observe that for all the benchmark test functions the CFA with adaptive local search obtains better or equivalent mean function value than its counterpart without adaptive local search.

Table 4. The mean best function value and the standard deviation obtained by CFA with or without adaptive local search.

Functions	No Adaptive Local Search		With Adaptive Local Search	
	Mean	Std	Mean	Std
Easom (2)	−0.99997	0.000020	−1.0	0.0
Shubert (2)	−185.779	0.963969	−186.723	0.016331
Rosenbrock (2)	0.0	0.0	0.0	0.0
Zakharov (2)	0.0	0.0	0.0	0.0
De Jong (3)	0.0	0.0	0.0	0.0
Shekel (4, 5)	−6.50369	2.613879	−8.09613	3.003044
Shekel (4, 7)	−7.40191	1.802712	−10.2611	0.440954
Shekel (4, 10)	−7.80353	1.033856	−10.3554	0.966638
Sphere (10)	0.0	0.0	0.0	0.0
Rosenbrock (10)	0.797320	1.594630	0.002511	0.009392
Rastrigin (10)	0.734548	0.827677	0.646338	0.800856
Griewank (10)	0.015448	0.011793	0.016450	0.009914
Zakharov (10)	0.0	0.0	0.0	0.0
Sphere (20)	0.0	0.0	0.0	0.0
Rosenbrock (20)	0.797828	1.594400	0.267050	0.994099
Rastrigin (20)	2.590780	1.391706	2.466102	1.765681
Griewank (20)	0.012054	0.013121	0.007213	0.012479
Zakharov (20)	0.0	0.0	0.0	0.0
Sphere (30)	0.0	0.0	0.0	0.0
Rosenbrock (30)	0.801375	1.599876	0.801094	1.598004
Rastrigin (30)	4.704743	1.842029	4.680690	2.632701
Griewank (30)	0.005663	0.007891	0.003364	0.007581
Zakharov (30)	0.0	0.0	0.0	0.0

4.2.3. Multi-Start Strategy for Swarm Rebuilding

Multi-start is a diversification strategy which terminates an ineffective trajectory search and reinitiates a new one from an under-explored region. Our CFA applies the adaptive multi-start strategy which monitors the performance of the incumbent firefly swarm. If the program stops improving the performance for a threshold number of consecutive iterations, part of the swarm is rebuilt by positioning some fireflies on diversified locations using the path-relinking technique. The performance stagnation threshold is made adaptive according to the result of the landscape analysis such that the frequency of the multi-start activation depends on the regional function profiles under exploration. It can be seen from Table 5 that the multi-start strategy effectively improves the mean best function value and the standard deviation obtained by CFA.

4.2.4. Responsive Strategies Based on Landscape Analysis

As noted in Section 3.5, the responsive strategies employed by the CFA are made adaptive based on the analysis of function landscape. The landscape analysis makes distinctions of two classic function forms: single-modal and multi-modal. The responsive strategies then vary the movement step size and the frequency of the local search and the multi-start activations to make the CFA search more effective. Table 6 lists the comparative performance of the CFA with or without performing the landscape analysis. It is noted that the version of the CFA without adaptive landscape analysis still embeds the local search and the multi-start procedures as its rudimentary components, which, however, are executed with fixed parameter values. From the tabulated result, we conclude that the landscape analysis can proliferate the performance gains possibly obtained by the local search and the multi-start strategies.

Table 5. The mean best function value and the standard deviation obtained by CFA with or without multi-start strategy.

Functions	No Multi-Start Rebuilding		With Multi-Start Rebuilding	
	Mean	Std	Mean	Std
Easom (2)	−1.0	0.0	−1.0	0.0
Shubert (2)	−186.03	0.708244	−186.723	0.016331
Rosenbrock (2)	0.0	0.0	0.0	0.0
Zakharov (2)	0.0	0.0	0.0	0.0
De Jong (3)	0.0	0.0	0.0	0.0
Shekel (4, 5)	−5.56783	3.517801	−8.09613	3.003044
Shekel (4, 7)	−9.12732	2.851160	−10.2611	0.440954
Shekel (4, 10)	−7.65202	3.724529	−10.3554	0.966638
Sphere (10)	0.0	0.0	0.0	0.0
Rosenbrock (10)	0.000006	0.000005	0.002511	0.009392
Rastrigin (10)	1.790926	1.294722	0.646338	0.800856
Griewank (10)	0.018083	0.015917	0.01645	0.009914
Zakharov (10)	0.0	0.0	0.0	0.0
Sphere (20)	0.0	0.0	0.0	0.0
Rosenbrock (20)	0.931281	1.685579	0.267050	0.994099
Rastrigin (20)	3.349695	1.807129	2.466102	1.765681
Griewank(20)	0.008904	0.011122	0.007213	0.012479
Zakharov (20)	0.0	0.0	0.0	0.0
Sphere (30)	0.0	0.0	0.0	0.0
Rosenbrock (30)	1.205757	1.838271	0.801094	1.598004
Rastrigin (30)	4.974795	1.973266	4.680690	2.632701
Griewank (30)	0.005735	0.011983	0.003364	0.007581
Zakharov (30)	0.000001	0.000004	0.0	0.0

4.2.5. Worst-Case Analysis

As the proposed CFA is a stochastic optimization algorithm, every single execution of the program may produce a different solution. It thus becomes very important to measure the solution variation of the worst case obtained by the CFA. We conduct the worst-case analysis on the three most challenging functions, namely Rosenbrock (30), Rastrigin (30), and Griewank (30) as follows. One thousand independent runs of the CFA program are performed. We record the worst function value (fitness) that could be obtained by allowing different number of repetitive runs in the experiment. It is seen from Figure 2 that the Rosenbrock (30) fitness value of less than 5 can be obtained with 99.7% confidence because three out of the one thousand runs report a fitness value greater than 5. It also indicates the worst fitness value we could obtain is no more than 5 if we can run the CFA program at least three times. Actually, there are two major components with the fitness distribution. One is close to 4 (a local optimum) with about 20% of the distribution, the other is near zero (the global optimum) with about 79% confidence. The worst-case analysis with Rastrigin (30) is illustrated in Figure 3. We observe a Gaussian-like distribution and the major component falls in the fitness value between 0 and 12. The Gaussian-like distribution provides a reliable guarantee that the mean performance value can be obtained with high confidence. Figure 4 shows the worst-case analysis for Griewank (30). The distribution has a long tail, and the major component concentrates at the high quality part. This situation manifests good properties that there is only one major outcome which is near the global optimum, and that a near-optimal solution can be obtained with a few runs in the worst case.

Table 6. The mean best function value and the standard deviation obtained by CFA with or without landscape analysis.

Functions	No Adaptive Landscape Analysis		With Adaptive Landscape Analysis	
	Mean	Std	Mean	Std
Easom (2)	−0.99997	0.000020	−1.0	0.0
Shubert (2)	−185.779	0.963969	−186.723	0.016331
Rosenbrock (2)	0.0	0.0	0.0	0.0
Zakharov (2)	0.0	0.0	0.0	0.0
De Jong (3)	0.0	0.0	0.0	0.0
Shekel (4, 5)	−6.50369	2.613879	−8.09613	3.003044
Shekel (4, 7)	−7.40191	1.802712	−10.2611	0.440954
Shekel (4, 10)	−7.80353	1.033856	−10.3554	0.966638
Sphere (10)	0.0	0.0	0.0	0.0
Rosenbrock (10)	0.797320	1.594630	0.002511	0.009392
Rastrigin (10)	0.734548	0.827677	0.646338	0.800856
Griewank (10)	0.015448	0.011793	0.016450	0.009914
Zakharov (10)	0.0	0.0	0.0	0.0
Sphere (20)	0.0	0.0	0.0	0.0
Rosenbrock (20)	0.797828	1.594400	0.267050	0.994099
Rastrigin (20)	2.590780	1.391706	2.466102	1.765681
Griewank (20)	0.012054	0.013121	0.007213	0.012479
Zakharov (20)	0.0	0.0	0.0	0.0
Sphere (30)	0.0	0.0	0.0	0.0
Rosenbrock (30)	0.801375	1.599876	0.801094	1.598004
Rastrigin (30)	4.704743	1.842029	4.680690	2.632701
Griewank (30)	0.005663	0.007891	0.003364	0.007581
Zakharov (30)	0.0	0.0	0.0	0.0

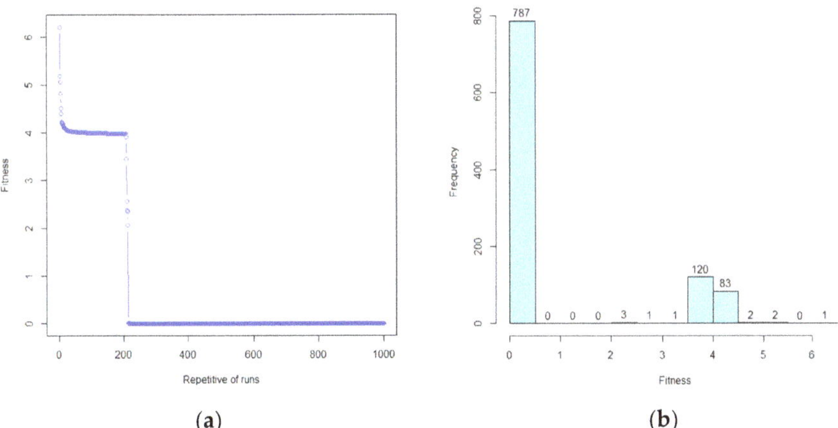

Figure 2. Worst-case analysis with the Rosenbrock (30) function. (**a**) The worst fitness obtained by the CFA program as the number of repetitive runs increases. (**b**) The number of program runs with which each performance level is reached.

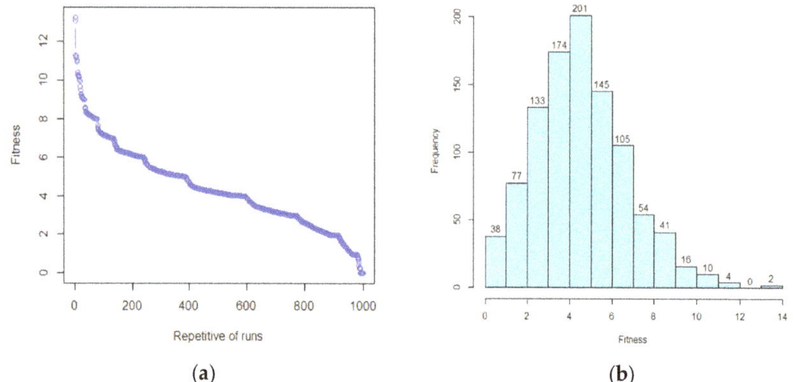

(a)

(b)

Figure 3. Worst-case analysis with the Rastrigin (30) function. (**a**) The worst fitness obtained by the CFA program as the number of repetitive runs increases. (**b**) The number of program runs with which each performance level is reached.

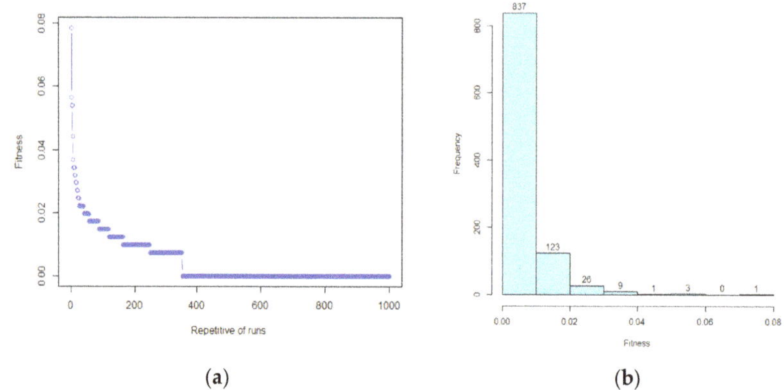

(a)

(b)

Figure 4. Worst-case analysis with the Griewank (30) function. (**a**) The worst fitness obtained by the CFA program as the number of repetitive runs increases. (**b**) The number of program runs with which each performance level is reached.

4.3. Performance Evaluation

In this section, the performance of the CFA is evaluated in three-fold. First, the CFA is compared against its counterparts, the GSO and the FA. Secondly, the CFA is compared to other kinds of metaheuristics. Thirdly, the performance of CFA is further justified on the CEC 2005 dataset. For the first two experiments, all compared algorithms were executed 30 times for each test function. For each run, the program can perform 160,000 FEs. However, for the third experiment, the evaluation criteria in the original paper are respected where the competing method is executed in 25 independent runs and in each run the method is evaluated at different numbers of consumed FEs. To compare the performance between two competing algorithms, we employ the performance index defined by Yin et al. [6] as follows. Given two competing algorithms, p and q, the performance merit of p against q on a test function is defined by the formula,

$$\text{Merit}(p, q) = (f_p - f^* + \varepsilon)/(f_q - f^* + \varepsilon) \tag{11}$$

where ε is a small constant equal to 5×10^{-7}, f_p and f_q are the mean best function values obtained by the competing algorithms p and q, and f* is the global minimum of the test function. As all the test functions involve minimization, we realize p outperforms q if Merit(p, q) < 1.0, p is inferior to q if Merit(p, q) > 1.0, and p and q perform equally well if Merit(p, q) = 1.0.

4.3.1. Comparison with CFA Counterparts

Our CFA enhances the GSO and the FA by incorporating the responsive strategies from the adaptive memory programming domain. Thus, it is important to validate our idea by comparing the performance of the CFA against its counterparts, the GSO and the FA. The results shown in the second to the fifth columns of Table 7 are the mean best function values obtained by the competing algorithms, and those in the last three columns correspond to the relative merit values. We observe that for the test functions with fewer than ten variables, the CFA has a unit merit or a merit value less than one with one order of magnitude, indicating that all the competing algorithms perform nearly equally well. For the test functions with ten or more variables, the merit value of the CFA in relation to the GSO and the FA becomes significantly less, implying a superior performance in favor of the CFA. The product of the merit values gives an overview of the comparative performance on all test function. It is seen at the bottom of Table 7 that the CFA has a merit product of 7.32×10^{-43} and 9.51×10^{-32} in relation to the GSO and the FA, respectively. We further compare the CFA to the better one of the best function values obtained by the GSO and the FA, giving a merit product of 8.99×10^{-29} as shown in the last column of Table 7. The result suggests that the CFA significantly outperforms its counterparts over the benchmark dataset, and that the CFA also beats a hybrid which takes the better one of the best function values obtained by the two counterparts.

Table 7. The performance comparison between the CFA and its counterparts.

Functions	(a) GSO	(b) FA	(c) Min (a, b)	(d) CFA	Merit (d, a)	Merit (d, b)	Merit (d, c)
Easom (2)	−0.999999	−0.999999	−0.999999	−1.0	1.0	1.0	1.0
Shubert (2)	−184.059055	−91.464488	−184.059055	−186.723	9.86×10^{-1}	4.90×10^{-1}	9.86×10^{-1}
Rosenbrock (2)	0.000007	0.0	0.0	0.0	6.50×10^{-2}	1.0	1.0
Zakharov (2)	0.0	0.0	0.0	0.0	1.0	1.0	1.0
De Jong (3)	0.0	0.0	0.0	0.0	1.0	1.0	1.0
Shekel (4, 5)	−6.883340	−3.339854	−6.883340	−8.09613	8.50×10^{-1}	4.13×10^{-1}	8.50×10^{-1}
Shekel (4, 7)	−6.624701	−2.593236	−6.624701	−10.2611	6.46×10^{-1}	2.53×10^{-1}	6.46×10^{-1}
Shekel (4, 10)	−7.770920	−2.211454	−7.770920	−10.3554	7.50×10^{-1}	2.14×10^{-1}	7.50×10^{-1}
Sphere (10)	0.021677	0.000045	0.000045	0.0	2.31×10^{-5}	1.11×10^{-2}	1.11×10^{-2}
Rosenbrock (10)	1.658934	1.070858	1.070858	0.002511	1.51×10^{-3}	2.35×10^{-3}	2.35×10^{-3}
Rastrigin (10)	2.499270	11.099298	2.499270	0.646338	2.59×10^{-1}	5.82×10^{-2}	2.59×10^{-1}
Griewank (10)	0.154192	17.666979	0.154192	0.01645	1.07×10^{-1}	9.31×10^{-4}	1.07×10^{-1}
Zakharov (10)	0.000330	0.000115	0.000115	0.0	1.51×10^{-3}	4.31×10^{-3}	4.31×10^{-3}
Sphere (20)	0.085764	0.000264	0.000264	0.0	5.83×10^{-6}	1.89×10^{-3}	1.89×10^{-3}
Rosenbrock (20)	8.473378	21.006502	8.473378	0.26705	3.15×10^{-2}	1.27×10^{-2}	3.15×10^{-2}
Rastrigin (20)	5.550878	32.150683	5.550878	2.466102	4.44×10^{-1}	7.67×10^{-2}	4.44×10^{-1}
Griewank (20)	0.553463	0.600160	0.553463	0.007213	1.30×10^{-2}	1.20×10^{-2}	1.30×10^{-2}
Zakharov (20)	0.002574	0.001387	0.001387	0.0	1.94×10^{-4}	3.60×10^{-4}	3.60×10^{-4}
Sphere (30)	0.151057	0.000761	0.000761	0.0	3.31×10^{-6}	6.57×10^{-4}	6.57×10^{-4}
Rosenbrock (30)	23.373028	31.808803	23.373028	0.801094	3.43×10^{-2}	2.52×10^{-2}	3.43×10^{-2}
Rastrigin (30)	9.135285	59.807848	9.135285	4.68069	5.12×10^{-1}	7.83×10^{-2}	5.12×10^{-1}
Griewank (30)	1.053057	0.010732	0.010732	0.003364	3.19×10^{-3}	3.13×10^{-1}	3.13×10^{-1}
Zakharov (30)	2.774662	0.008677	0.008677	0.000001	4.98×10^{-7}	1.59×10^{-4}	1.59×10^{-4}
Merit Product					$\mathbf{7.32 \times 10^{-43}}$	$\mathbf{9.51 \times 10^{-32}}$	$\mathbf{8.99 \times 10^{-29}}$

We further verify the online performance advantage of our CFA against its counterparts by a statistical test suggested in Taillard [28]. As the best objective value produced by a metaheuristic approach is non-deterministic, we model the result obtained from multiple runs of method A (and method B) as a random variable X_a (X_b) and we want to testify the confidence regarding that X_a is less than X_b. A classic statistical test based on the central limit theorem for comparing two proportions is to approximate the mean $X_a - X_b$ as a normal distribution if the collected number of samples is sufficiently large. Therefore, 30 independent runs of each competing algorithm were conducted. For each run, the online best function value obtained at a particular FE, say e, is a sample for $X_a(e)$, the random variable for the result obtained by method A at FE e. We tally the samples at every instance of FEs during the whole duration of executing the algorithm. By examining the mean curve of $X_a(e)$ and $X_b(e)$ as e increases during the evolution, we can testify if method A well outperforms method B. For clear illustration, the boundary and mean curves over the 30 runs are plotted. Figure 5a shows the online performance analysis with 95% confidence interval for the Rosenbrock (30) function. It is seen that the 95% confidence interval of the best function value obtained by the three competing algorithms (GSO, FA, and CFA) converges with various speeds. Our CFA converges at a much faster speed and reach towards a better function value than its two counterparts. The FA is the second-best performer followed by the GSO. To investigate the detailed performance during the second half duration of the execution, we enlarge the plot for this period as shown in Figure 5b. It can be seen that the CFA significantly outperforms the other two algorithms with 95% confidence level during the second half execution duration. We also found that the GSO performs better than the FA after 80,000 FEs, although the GSO may not beat the FA at the early stage of the execution as previously noted. The online performance comparison with 95% confidence level for the Rastrigin (30) function is shown in Figure 6. We observe that during the whole execution period the CFA significantly outperforms the GSO and the FA. The FA performs better than the GSO when the allowed number of consumed FEs is less than 20,000, but the FA is far surpassed by the GSO if more FEs are allowed. Figure 7 shows the online performance variation with 95% confidence level for the Griewank (30) function. Again, the CFA is the best performer among the three algorithms throughout the whole execution duration. However, we see a phenomenon differing from those for the two previous test functions with the GSO and the FA. The GSO and the FA performs about equally well before consuming 50,000 FEs, although the former is a more stable performer because it has a shorter confidence interval. However, after this critical execution period, the FA becomes very effective both in the convergence speed and the function value. As shown in Figure 7b, the FA significantly surpasses the GSO and reaches a comparative performance with the CFA.

4.3.2. Comparison against Other Metaheuristics

We now compare the CFA against other metaheuristics inspired by different nature metaphors, the PSO, the GA, and the cyber swarm algorithm (CSA). The compared PSO is the constriction factor version proposed by Clerc and Kennedy [29] which has been shown to be one of the best PSO implementations. The implemented GA employs real-value chromosome coding, tournament selection (with $k = 2$, i.e., two competitors in each instance of selection), arithmetic crossover, and Gaussian mutation. The GA is generational without population gap, i.e., the whole parent population is replaced by the offspring population. The implementation and parameter setting of CSA follow the original paper [6]. All the compared algorithms have the same population size of 60 individuals, and are executed until consuming 160,000 FEs. Table 8 tabulates the mean best function value obtained by the compared algorithms over 30 runs and the merit value among the competitors. For the comparison of the CFA against the PSO and the GA, we observe that the CFA well surpasses the other two algorithms on most of the benchmark functions. The merit product in relation to the PSO and the GA is 8.80×10^{-21} and 2.11×10^{-36}, respectively. When we compare the CFA to the CSA, the merit product is 1.94×10^{18}. The result seems to suggest that the CSA performs better on the dataset. However, if we take a closer look, the CSA is very effective in solving small functions with less than ten variables, thus CSA gives significantly greater merits for these functions. For the test functions with ten or more variables, the merit value turns to be in favor to the CFA, disclosing that the CFA is more effective than CSA in tackling larger-sized functions. It is worth noting that both CFA and the CSA take advantage of the features contained in the AMP domain, and the two algorithms extremely outperform the other compared algorithms in our experiments. This phenomenon discloses the potential of future research in the direction of marrying the AMP with other types of metaheuristics.

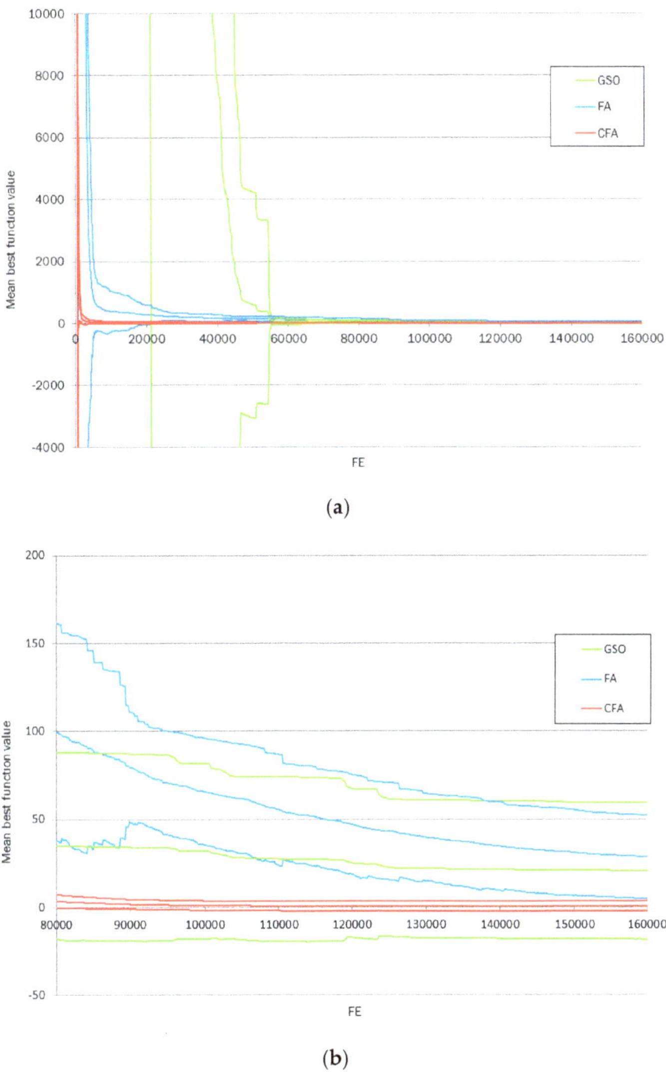

(a)

(b)

Figure 5. Online performance analysis with 95% confidence interval for the Rosenbrock (30) function. (**a**) The convergence of the best function value during the whole duration of the execution. (**b**) The convergence of the best function value during the second half duration of the execution.

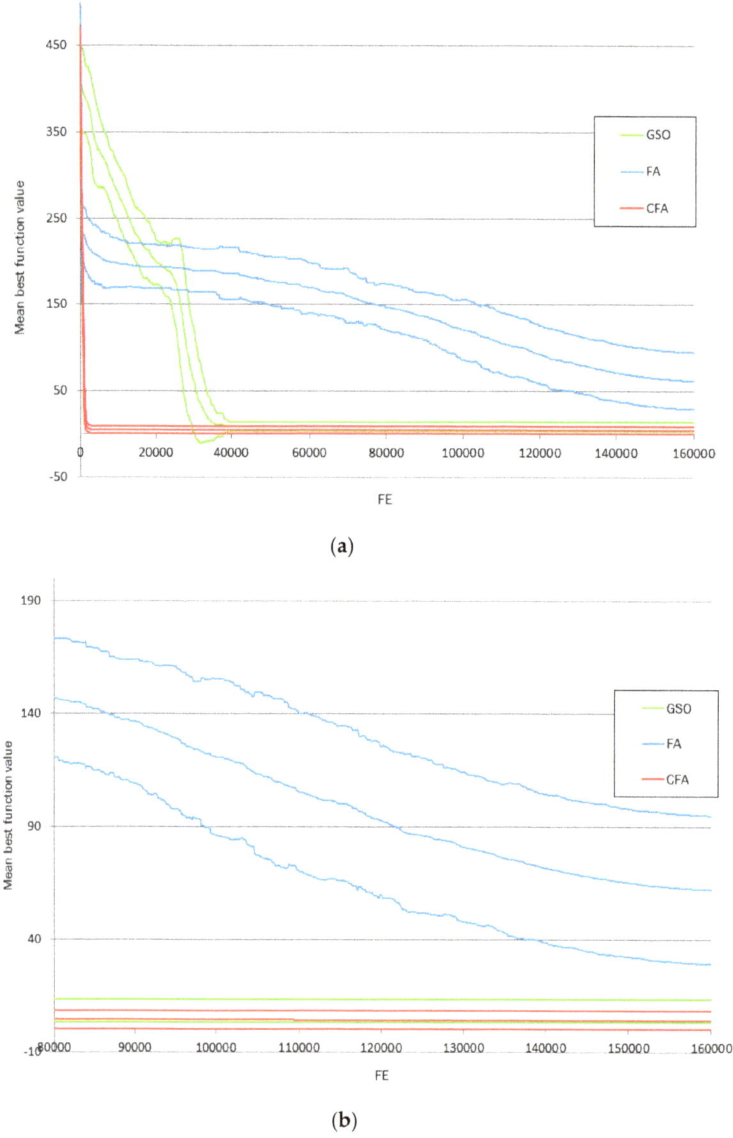

(**a**)

(**b**)

Figure 6. Online performance analysis with 95% confidence interval for the Rastrigin (30) function. (**a**) The convergence of the best function value during the whole duration of the execution. (**b**) The convergence of the best function value during the second half duration of the execution.

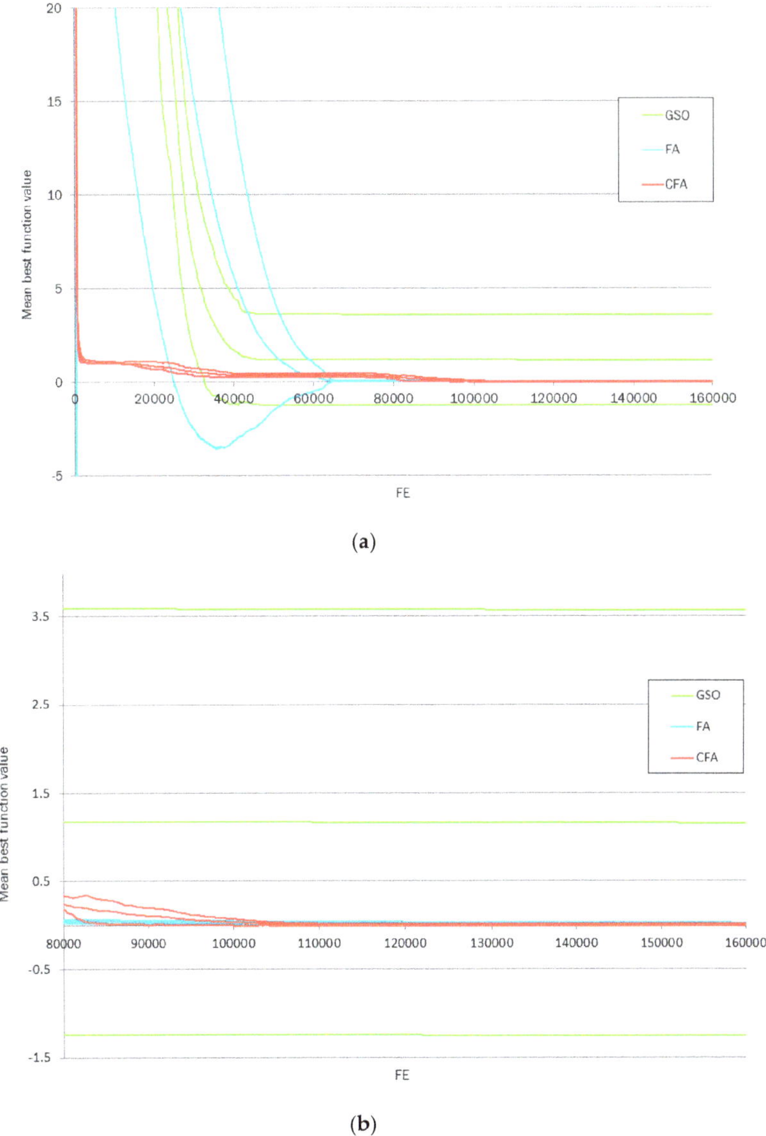

(**a**)

(**b**)

Figure 7. Online performance analysis with 95% confidence interval for the Griewank (30) function. (**a**) The convergence of the best function value during the whole duration of the execution. (**b**) The convergence of the best function value during the second half duration of the execution.

Table 8. The performance comparison between the CFA and other metaheuristics.

Functions	(a) PSO	(b) GA	(c) CSA	(d) CFA	Merit (d, a)	Merit (d, b)	Merit (d, c)
Easom (2)	−1.0	−1.0	−1.0	−1.0	1.0	1.0	1.0
Shubert (2)	−186.7309	−186.7309	−186.7309	−186.7232	1.54×10^4	1.54×10^4	1.54×10^4
Rosenbrock (2)	0.0	0.1015	0.0	0.0	1.0	4.93×10^{-6}	1.0
Zakharov (2)	0.0	0.0	0.0	0.0	1.0	1.0	1.0
De Jong (3)	0.0	0.0	0.0	0.0	1.0	1.0	1.0
Shekel (4, 5)	−6.6329	−10.0535	−10.1532	−8.0961	5.84×10^{-1}	2.06×10^1	4.11×10^6
Shekel (4, 7)	−8.0176	−10.0637	−10.4029	−10.2611	5.94×10^{-2}	4.18×10^{-1}	2.84×10^5
Shekel (4, 10)	−7.4195	−10.0750	−10.5364	−10.3554	5.81×10^{-2}	3.92×10^{-1}	3.62×10^5
Sphere (10)	0.0	0.0009	0.0	0.0	1.0	5.55×10^{-4}	1.0
Rosenbrock (10)	0.4727	8.1009	0.1595	0.0025	5.29×10^{-3}	3.09×10^{-4}	1.57×10^{-2}
Rastrigin (10)	6.4672	0.0004	0.7464	0.6463	9.99×10^{-2}	1.61×10^3	8.66×10^{-1}
Griewank (10)	0.0644	0.0493	0.0474	0.0164	2.55×10^{-1}	3.33×10^{-1}	3.46×10^{-1}
Zakharov (10)	0.0	0.1809	0.0	0.0	1.0	2.76×10^{-6}	1.0
Sphere (20)	0.0	0.0069	0.0	0.0	1.0	7.25×10^{-5}	1.0
Rosenbrock (20)	0.3992	9.0326	0.4788	0.2671	6.69×10^{-1}	2.96×10^{-2}	5.58×10^{-1}
Rastrigin (20)	18.7052	0.0036	6.8868	2.4661	1.32×10^{-1}	6.85×10^2	3.58×10^{-1}
Griewank (20)	0.0227	0.0533	0.0128	0.0072	3.17×10^{-1}	1.35×10^{-1}	5.63×10^{-1}
Zakharov (20)	2.683	37.1846	0.0	0.0	1.86×10^{-7}	1.34×10^{-8}	1.0
Sphere (30)	0.0	0.0226	0.0	0.0	1.0	2.21×10^{-5}	1.0
Rosenbrock (30)	9.2089	83.0118	0.3627	0.8011	8.70×10^{-2}	9.65×10^{-3}	2.21
Rastrigin (30)	33.9281	0.0115	11.9425	4.6807	1.38×10^{-1}	4.07×10^2	3.92×10^{-1}
Griewank (30)	0.0093	0.0893	0.0052	0.0034	3.66×10^{-1}	3.81×10^{-2}	6.54×10^{-1}
Zakharov (30)	5.4347	136.7129	0.0	0.0	9.20×10^{-8}	3.66×10^{-9}	1.0
Merit Product					$\mathbf{8.80 \times 10^{-21}}$	$\mathbf{2.11 \times 10^{-36}}$	$\mathbf{1.94 \times 10^{18}}$

To compare the CFA with the state-of-the-art variants of FA, we quote the results (mean objective value over 30 runs) from the original paper LFA [9], VESSFA [10], WFA [11], CLFA [12], FAtidal [13], and GDAFA [14]. The best mean objective value for each function obtained by all compared methods is printed in bold. As can be seen in Table 9, our CFA wins the most times as obtaining the best mean objective value among all. GDAFA seems to possess better performance as the dimensionality increases. Both CFA and GDAFA can gain an objective value very close to the optimum, while the other competing methods may produce an objective value far away from the optimum in some challenging functions.

Table 9. The performance comparison between the CFA and the state-of-the-art variants of FA.

Functions	LFA	FAtidal	VSSFA	GDAFA	WFA	CLFA	CFA
Sphere (2)	0.043	**0.0**	–	–	–	–	0.0
Rosenbrock (2)	1.34	0.0076	–	–	–	–	0.0
Zakharov (2)	0.4950	**0.0**	–	–	–	–	0.0
Sphere (10)	17.5	**0.0**	0.0	0.0	0.0690	0.0	0.0
Rosenbrock (10)	3.15×10^4	9.93	–	–	–	–	**0.0025**
Rastrigin (10)	85.5	7.20	8.6769	**0.0**	10.6039	11.4284	0.6463
Griewank (10)	0.0069	–	0.0	0.0	0.0075	0.0	0.0164
Zakharov (10)	158.0	**0.0**	–	–	–	–	0.0
Sphere (30)	14.0818	–	17.0501	1.61×10^{-5}	0.2373	0.1942	**0.0**
Rastrigin (30)	38.3584	–	133.9519	**0.0619**	42.7987	91.6579	4.6807
Griewank (30)	0.4365	–	0.5380	$\mathbf{3.65 \times 10^{-6}}$	0.0123	0.0099	0.0034

4.3.3. Comparison on the CEC 2005 Dataset

To further justify the performance of CFA, we compare CFA with the investigated methods reported in [26] on 12 CEC 2005 benchmark functions [30]. The IEEE CEC Repository [27] provides fruitful benchmark datasets for optimization problems with various purposes such as unconstrained, constrained, and multi-objective optimization. The CEC 2005 dataset is designed for unconstrained real-parameter optimization which is addressed in this paper. We selected 12 CEC 2005 functions which are very challenging and have never been solved to optimal by any known methods [26]. We executed all compared algorithms with the same evaluation criteria and parameter settings as described in the original paper [26]. Each algorithm is executed for 25 independent runs on each test function with $n = 10$ and 30, respectively. All compared methods are executed by being allowed to consume 1000, 10,000, and 100,000 FEs.

We adopt the GAP performance measure proposed in the original paper [26] and it is defined as GAP = |f − f*| where f is the function value obtained by an evaluated method and f* is the global optimum value of the test function. Table 10 shows the mean minimum (Min.) and average (Avg.) of GAP and Merit of all compared methods over the 12 functions for $n = 10$. We observe that our CFA is less exploitative in small size CEC problems than the leading methods such as G-CMA-ES and L-CMA-ES, both of which are based on the covariance matrix adaptation evolution strategy (CMA-ES) [31] which updates the covariance of the multivariate distribution to better handle the dependency between variables. Though the CFA is less competitive in the mean Avg. GAP, it can deliver a quality Min. out of 25 independent runs. It suggests that the CFA can be executed multiple times and output the best value from those runs when tackling small size yet complex problems. This phenomenon is also revealed in the geometric mean of the merits (GMM). The GMM of the Min. function value is 1.099, 0.995, and 0.878 at 1000, 10,000, and 100,000 FEs, while the GMM of the Avg. function value gradually deteriorates from 0.963, 1.167 to 1.211 at the same FEs.

Table 10. Min./Avg. GAP and Merit over the CEC dataset with $n = 10$.

FEs	GAP			Merit		
	1000	10,000	100,000	1000	10,000	100,000
	Min./Avg.	Min./Avg.	Min./Avg.	Min./Avg.	Min./Avg.	Min./Avg.
CFA	572.3/751.2	317.6/554.4	212.4/441.1	1.0/1.0	1.0/1.0	1.0/1.0
CSA	382.0/665.9	291.3/506.6	234.5/432.6	1.498/1.128	1.090/1.094	0.906/1.020
STS	616.1/759.4	348.9/576.6	198.3/413.4	0.929/0.989	0.910/0.961	1.071/1.067
G-CMA-ES	269.7/542.0	260.0/419.4	256.0/265.3	2.122/1.386	1.222/1.322	0.830/1.663
EDA	669.9/1059.1	287.1/335.1	269.4/300.6	0.854/0.709	1.106/1.654	0.788/1.467
BLX-MA	456.7/711.1	315.5/445.1	306.2/430.1	1.253/1.056	1.007/1.246	0.694/1.026
SPC-PNX	621.7/750.3	279.6/391.0	206.0/309.9	0.921/1.001	1.136/1.418	1.031/1.423
BLX-GL50	676.0/716.3	272.8/341.0	257.2/319.0	0.847/1.049	1.164/1.626	0.826/1.383
L-CMA-ES	289.0/825.7	225.9/655.8	202.7/411.1	1.980/0.910	1.406/0.845	1.048/1.073
DE	715.4/914.1	396.7/492.4	228.8/272.0	0.800/0.822	0.801/1.126	0.928/1.622
K-PCX	671.0/968.5	488.0/564.4	257.4/475.6	0.853/0.776	0.651/0.982	0.825/0.927
Co-EVO	672.6/799.0	437.5/623.5	268.3/465.4	0.851/0.940	0.726/0.889	0.792/0.948
Merit Product				2.833/0.663	0.949/5.493	0.238/8.195
Geometric Mean				1.099/0.963	0.995/1.167	0.878/1.211

As for high-dimensional and complex CEC problems with $n = 30$, the mean Min. and Avg. of GAP and Merit of all compared methods are tabulated in Table 11. It is seen from the GMM that our CFA is compared favorably to the other methods in both Min. and Avg. function value at all FEs check points. The prevailing exploration search conducted by CFA is due to its elements of AMP-responsive strategies, which are more effective when the problem is more complex and is presented in higher-dimensional space. The G-CMA-ES is again the best method since it excels in terms of GAP in most cases as compared to the other competing methods. It is worth further studying the possibility of including the CMA technique into the CFA to resolve the dependency between variables.

Table 11. Min./Avg. GAP and Merit over the CEC dataset with $n = 30$.

FEs	GAP			Merit		
	1000	10,000	100,000	1000	10,000	100,000
	Min./Avg.	Min./Avg.	Min./Avg.	Min./Avg.	Min./Avg.	Min./Avg.
CFA	792.5/1033.1	450.8/778.6	418.7/602.5	1.0/1.0	1.0/1.0	1.0/1.0
CSA	629.2/785.0	454.4/647.8	420.6/578.2	1.260/1.316	0.992/1.202	0.995/1.042
STS	829.3/957.0	614.9/747.3	431.3/540.3	0.956/1.080	0.733/1.042	0.971/1.115
G-CMA-ES	570.3/658.4	414.4/526.8	405.7/493.0	1.390/1.569	1.088/1.478	1.032/1.222
EDA	39,742/63,491	11,951/26,418	653.6/934.7	0.020/0.016	0.038/0.029	0.641/0.645
BLX-MA	792.9/1198.7	443.9/502.4	410.7/457.2	0.999/0.862	1.016/1.550	1.019/1.318
SPC-PNX	29,793/74,050	637.6/850.1	414.8/430.0	0.027/0.014	0.707/0.916	1.009/1.401
BLX-GL50	8545.4/20,008	474.8/545.9	433.0/507.5	0.093/0.052	0.949/1.426	0.967/1.187
L-CMA-ES	790.8/1009.8	447.6/722.6	404.6/617.0	1.002/1.023	1.007/1.077	1.035/0.976
DE	3473.3/14,461	726.0/781.8	558.7/592.0	0.228/0.071	0.621/0.996	0.749/1.018
K-PCX	27,749/108,623	27,719/108,602	866.1/2257.2	0.029/0.010	0.016/0.007	0.483/0.267
Co-EVO	908.5/1025.8	7496/822.0	625.3/734.5	0.872/1.007	0.060/0.947	0.670/0.820
Merit Product				$4.7 \times 10^{-7}/2.0 \times 10^{-8}$	$1.2 \times 10^{-5}/8.0 \times 10^{-4}$	$1.5 \times 10^{-1}/4.3 \times 10^{-1}$
Geometric Mean				0.266/0.195	0.358/0.523	0.846/0.927

5. Concluding Remarks and Future Research

We have proposed the CFA which is a more effective form of the GSO and the FA in global optimization. The CFA incorporates several AMP strategies including multiple guiding solutions, pattern search as local improvement method, solution set rebuilding in a multi-start search template, and the responsive strategies. The experimental result on benchmark functions for global optimization has shown that the CFA performs significantly better in terms of both solution quality and robustness than the GSO, FA, and several state-of-the-art metaheuristic methods as demonstrated in our statistical analyses and comprehensive experiments. It is worth noting that it is certain a sophisticated method such as CFA incorporating advanced components will pose higher computational complexity in the computation iteration than an algorithm which does not. However, as metaheuristic approaches are computational ones which can stop at any computation iteration and output the best-so-far result. We conduct a fair performance comparison between two metaheuristic approaches at the same number of fitness FE instead of using the evolution iterations. All our experiments follow this fashion.

Our findings strengthen the motivations for marrying the approaches selected from each of the metaheuristic dichotomies, respectively. The CFA template gives general ideas for creating this sort of effective hybrid metaheuristics. Inspired by the promising result of the CFA, it is worthy of investigating the possibility of the application of the CFA template to other metaheuristic approaches with various problem domains for future research.

Author Contributions: Conceptualization, P.-Y.Y.; methodology, P.-Y.Y. and P.-Y.C.; software, P.-Y.C.; validation, Y.-C.W. and R.-F.D.; writing—original draft preparation, P.-Y.Y.; writing—review and editing, P.-Y.Y., Y.-C.W. and R.-F.D.; visualization, P.-Y.C.; funding acquisition, P.-Y.Y. All authors have read and agreed to the published version of the manuscript.

Funding: This research was funded by MOST Taiwan, grant numbers 107-2410-H-260-015-MY3. The APC was funded by 107-2410-H-260-015-MY3.

Conflicts of Interest: The authors declare no conflict of interest. The funders had no role in the design of the study; in the collection, analyses, or interpretation of data; in the writing of the manuscript, or in the decision to publish the results.

Nomenclature

The list of the acronyms referenced in this paper is tabulated as follows.

EA	evolutionary algorithm
GSO	glowworm swarm optimization
FA	firefly algorithm
AMP	adaptive memory programming
CSA	cyber swarm algorithm
CFA	Cyber Firefly Algorithm
GRASP	greedy randomized adaptive search procedures
GA	genetic algorithm
PSO	particle swarm optimization
VNS	variable neighborhood search
SS	scatter search
SS/PR	path relinking
FDC	fitness distance correlation
FE	function evaluations
CMA-ES	covariance matrix adaptation evolution strategy
GMM	geometric mean of the merits

References

1. Yin, P.Y. *Towards more effective metaheuristic computing, In Modeling, Analysis, and Applications in Metaheuristic Computing: Advancements and Trends*; IGI-Global Publishing: Hershey, PA, USA, 2012.
2. Talbi, E.G.; Bachelet, V. COSEARCH: A parallel cooperative metaheuristic. *J. Math. Model. Algorithms* **2006**, *5*, 5–22. [CrossRef]
3. Shen, Q.; Shi, W.M.; Kong, W. Hybrid particle swarm optimization and tabu search approach for selecting genes for tumor classification using gene expression data. *Comput. Biol. Chem.* **2008**, *32*, 52–59. [CrossRef] [PubMed]
4. Marinakis, Y.; Marinaki, M.; Doumpos, M.; Matsatsinis, N.F.; Zopounidis, C. A hybrid ACO-GRASP algorithm for clustering analysis. *Ann. Oper. Res.* **2011**, *188*, 343–358. [CrossRef]
5. Fuksz, L.; Pop, P.C. A hybrid genetic algorithm with variable neighborhood search approach to the number partitioning problem. *Lect. Notes Comput. Sci.* **2013**, *8073*, 649–658.
6. Yin, P.Y.; Glover, F.; Laguna, M.; Zhu, J.S. Cyber swarm algorithms: Improving particle swarm optimization using adaptive memory strategies. *Eur. J. Oper. Res.* **2010**, *201*, 377–389. [CrossRef]
7. Krishnanand, K.N.; Ghose, D. Detection of multiple source locations using a glowworm metaphor with applications to collective robotics. In Proceedings of the IEEE Swarm Intelligence Symposium, Pasadena, CA, USA, 8–10 June 2005; pp. 84–91.
8. Yang, X.S. Firefly algorithm. *Nat. Inspired Metaheuristic Algorithms* **2008**, *20*, 79–90.
9. Yang, X.S. Firefly algorithm, levy flights and global optimization. In *Research and Development in Intelligent Systems XXVI*; Springer: London, UK, 2010; pp. 209–218.
10. Yu, S.; Zhu, S.; Ma, Y.; Mao, D. A variable step size firefly algorithm for numerical optimization. *Appl. Math. Comput.* **2015**, *263*, 214–220. [CrossRef]
11. Zhu, Q.G.; Xiao, Y.K.; Chen, W.D.; Ni, C.X.; Chen, Y. Research on the improved mobile robot localization approach based on firefly algorithm. *Chin. J. Sci. Instrum.* **2016**, *37*, 323–329.
12. Kaveh, A.; Javadi, S.M. Chaos-based firefly algorithms for optimization of cyclically large-size braced steel domes with multiple frequency constraints. *Comput. Struct.* **2019**, *214*, 28–39. [CrossRef]
13. Yelghi, A.; Köse, C. A modified firefly algorithm for global minimum optimization. *Appl. Soft Comput.* **2018**, *62*, 29–44. [CrossRef]
14. Liu, J.; Mao, Y.; Liu, X.; Li, Y. A dynamic adaptive firefly algorithm with globally orientation. *Math. Comput. Simul.* **2020**, *174*, 76–101. [CrossRef]
15. Wang, J.; Song, F.; Yin, A.; Chen, H. Firefly algorithm based on dynamic step change strategy. In *Machine Learning for Cyber Security*; Chen, X., Yan, H., Yan, Q., Zhang, X., Eds.; Lecture Notes in Computer Science 12487; Springer: Cham, Switzerland, 2020. [CrossRef]
16. Glover, F. Tabu search and adaptive memory programming—Advances, applications and challenges. In *Interfaces in Computer Science and Operations Research*; Kluwer Academic Publishers: London UK, 1996; pp. 1–75.
17. Glover, F. A template for scatter search and path relinking. *Lect. Notes Comput. Sci.* **1998**, *1363*, 13–54.
18. Laguna, M.; Marti, R. *Scatter Search: Methodology and Implementation in C*; Kluwer Academic Publishers: London, UK, 2003.
19. Chen, X.S.; Ong, Y.S.; Lim, M.H.; Tan, K.C. A Multi-Facet Survey on Memetic Computation. *IEEE Trans. Evol. Comput.* **2011**, *15*, 591–607. [CrossRef]
20. Feo, T.A.; Resende, M.G.C. Greedy randomized adaptive search procedures. *J. Glob. Optim.* **1995**, *6*, 109–133. [CrossRef]
21. Hooke, R.; Jeeves, T.A. Direct search solution of numerical and statistical problems. *J. Assoc. Comput. Mach.* **1961**, *8*, 212–229. [CrossRef]
22. Dolan, E.D.; Lewis, R.M.; Torczon, V.J. On the local convergence of pattern search. *Siam J. Optim.* **2003**, *14*, 567–583. [CrossRef]
23. Jones, T.; Forrest, S. Fitness distance correlation as a measure of problem difficulty for genetic algorithms. In Proceedings of the International Conference on Genetic Algorithms, Morgan Laufman, Santa Fe, NM, USA, 15–19 July 1995; pp. 184–192.
24. Hedar, A.R.; Fukushima, M. Tabu search directed by direct search methods for nonlinear global optimization. *Eur. J. Oper. Res.* **2006**, *170*, 329–349. [CrossRef]

25. Hirsch, M.J.; Meneses, C.N.; Pardalos, P.M.; Resende, M.G.C. Global optimization by continuous GRASP. *Optim. Lett.* **2007**, *1*, 201–212. [CrossRef]
26. Duarte, A.; Marti, R.; Glover, F.; Gortazar, F. Hybrid scatter-tabu search for unconstrained global optimization. *Ann. Oper. Res.* **2011**, *183*, 95–123. [CrossRef]
27. Al-Roomi, A.R. *IEEE Congresses on Evolutionary Computation Repository*; Dalhousie University, Electrical and Computer Engineering: Halifax, NS, Canada, 2015; Available online: https://www.al-roomi.org/benchmarks/cec-database (accessed on 15 November 2020).
28. Taillard, E.D.; Waelti, P.; Zuber, J. Few statistical tests for proportions comparison. *Eur. J. Oper. Res.* **2008**, *185*, 1336–1350. [CrossRef]
29. Clerc, M.; Kennedy, J. The particle swarm explosion, stability, and convergence in a multidimensional complex space. *IEEE Trans. Evol. Comput.* **2002**, *6*, 58–73. [CrossRef]
30. Suganthan, P.N.; Hansen, N.; Liang, J.J.; Deb, K.; Chen, Y.P.; Auger, A.; Tiwari, S. *Problem Definitions and Evaluation Criteria for the CEC 2005 Special Session on Real-Parameter Optimization*; Technical Report; Nanyang Technology University of Singapore: Singapore, 2005.
31. Hansen, N. The CMA evolution strategy: A comparing review. In *Towards a New Evolutionary Computation. Advances on Estimation of Distribution Algorithms*; Springer: Berlin/Heidelberg, Germany, 2006; pp. 1769–1776.

Publisher's Note: MDPI stays neutral with regard to jurisdictional claims in published maps and institutional affiliations.

applied sciences

MDPI

Article

Search Patterns Based on Trajectories Extracted from the Response of Second-Order Systems

Erik Cuevas [1,*], Héctor Becerra [1], Héctor Escobar [1], Alberto Luque-Chang [1], Marco Pérez [1], Heba F. Eid [2] and Mario Jiménez [1]

[1] Departamento de Electrónica, Universidad de Guadalajara, CUCEI Av. Revolución 1500, Guadalajara 44430, Mexico; hectorg.becerra@academicos.udg.mx (H.B.); hector.11294@gmail.com (H.E.); alberto.lchang@academicos.udg.mx (A.L.-C.); marco.perez@cucei.udg.mx (M.P.); mario.jimenez@academicos.udg.mx (M.J.)
[2] Faculty of Science, Al-Azhar University, Cairo 11651, Egypt; heba.fathy@azhar.edu.eg
* Correspondence: erik.cuevas@cucei.udg.mx

Abstract: Recently, several new metaheuristic schemes have been introduced in the literature. Although all these approaches consider very different phenomena as metaphors, the search patterns used to explore the search space are very similar. On the other hand, second-order systems are models that present different temporal behaviors depending on the value of their parameters. Such temporal behaviors can be conceived as search patterns with multiple behaviors and simple configurations. In this paper, a set of new search patterns are introduced to explore the search space efficiently. They emulate the response of a second-order system. The proposed set of search patterns have been integrated as a complete search strategy, called Second-Order Algorithm (SOA), to obtain the global solution of complex optimization problems. To analyze the performance of the proposed scheme, it has been compared in a set of representative optimization problems, including multimodal, unimodal, and hybrid benchmark formulations. Numerical results demonstrate that the proposed SOA method exhibits remarkable performance in terms of accuracy and high convergence rates.

Keywords: metaheuristic methods; search patterns; second-order systems; evolutionary methods

Citation: Cuevas, E.; Becerra, H.; Escobar, H.; Luque-Chang, A.; Pérez, M.; Eid, H.F.; Jiménez, M. Search Patterns Based on Trajectories Extracted from the Response of Second-Order Systems. *Appl. Sci.* **2021**, *11*, 3430. https://doi.org/10.3390/app11083430

Academic Editor: Juan A. Gómez-Pulido

Received: 6 March 2021
Accepted: 1 April 2021
Published: 12 April 2021

Publisher's Note: MDPI stays neutral with regard to jurisdictional claims in published maps and institutional affiliations.

1. Introduction

Metaheuristic algorithms refer to generic optimization schemes that emulate the operation of different natural or social processes. In metaheuristic approaches, the optimization strategy is performed by a set of search agents. Each agent maintains a possible solution to the optimization problem, and is initially produced by considering a random feasible solution. An objective function determines the quality of the solution of each agent. By using the values of the objective function, at each iteration, the position of the search agents is modified, employing a set of search patterns that regulate their movements within the search space. Such search patterns are abstract models inspired by natural or social processes [1]. These steps are repeated until a stop criterion is reached. Metaheuristic schemes have confirmed their supremacy in diverse real-world applications in circumstances where classical methods cannot be adopted.

Essentially, a clear classification of metaheuristic methods does not exist. Despite this, several categories have been proposed that considered different criteria, such as a source of inspiration, type of operators or cooperation among the agents. In relation to inspiration, nature-inspired metaheuristic algorithms are classified into three categories: Evolution-based, swarm-based, and physics-based. Evolution-based approaches correspond to the most consolidate search strategies that use evolution elements as operators to produce search patterns. Consequently, operations, such as reproduction, mutation, recombination, and selection are used to generate search patterns during their operations. The most representative examples of evolution-based techniques, include Evolutionary Strategies

(ES) [2–4], Genetic Algorithms (GA) [5], Differential Evolution (DE) [6] and Self-Adaptative Differential Evolution (JADE) [7]. Swarm-inspired techniques use behavioral schemes extracted from the collaborative interaction of different animals or species of insects to produce a search strategy. Recently, a high number of swarm-based approaches have been published in the literature. Among the most popular swarm-inspired approaches, include the Crow Search Algorithm (CSA) [8], Artificial Bee Colony (ABC) [9], Particle Swarm Optimization (PSO) algorithm [10–12], Firefly Algorithm (FA) [13,14], Cuckoo Search (CS) [15], Bat Algorithm (BA) [16], Gray Wolf Optimizer (GWO) [17], Moth-flame optimization algorithm (MFO) [18] to name a few. Metaheuristic algorithms that consider the physics-based scheme use simplified physical models to produce search patterns for their agents. Some examples of the most representative physics-based techniques involve the States of Matter Search (SMS) [19,20], the Simulated Annealing (SA) algorithm [21–23], the Gravitational Search Algorithm (GSA) [24], the Water Cycle Algorithm (WCA) [25], the Big Bang-Big Crunch (BB-BC) [26] and Electromagnetism-like Mechanism (EM) [27]. Figure 1 visually exhibits the taxonomy of the metaheuristic classification. Although all these approaches consider very different phenomena as metaphors, the search patterns used to explore the search space are exclusively based on spiral elements or attraction models [10–17,28]. Under such conditions, the design of many metaheuristic methods refers to configuring a recycled search pattern that has been demonstrated to be successful in previous approaches for generating new optimization schemes through a marginal modification.

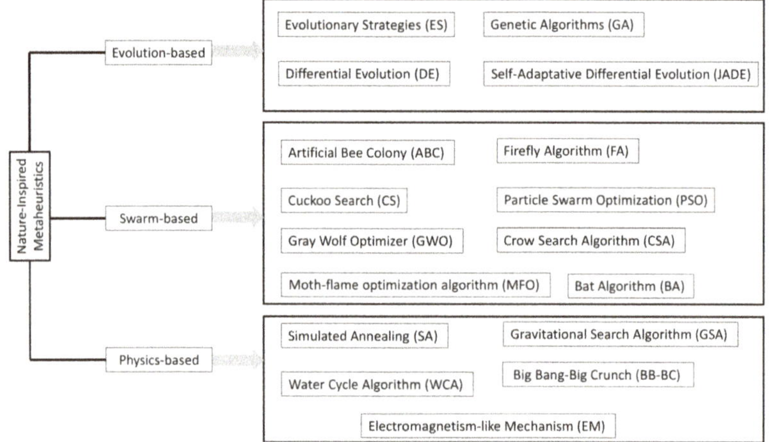

Figure 1. Visual taxonomy of the nature-inspired metaheuristic schemes.

On the other hand, the order of a differential equation refers to the highest degree of derivative considered in the model. Therefore, a model whose input-output formulation is a second-order differential equation is known as a second-order system [29]. One of the main elements that make a second-order model important is its ability to present very different behaviors, depending on the configuration of its parameters. Through its different behaviors, such as oscillatory, underdamped, or overdamped, a second-order system can exhibit distinct temporal responses [30]. Such behaviors can be observed as search trajectories under the perspective of metaheuristic schemes. Therefore, with second-order systems, it is possible to produce oscillatory movements within a certain region or build complex search patterns around different points or sections of the search space.

In this paper, a set of new search patterns are introduced to explore the search space efficiently. They emulate the response of a second-order system. The proposed set of search patterns have been integrated as a complete search strategy, called Second-Order Algorithm (SOA), to obtain the global solution of complex optimization problems. To analyze the performance of the proposed scheme, it has been compared in a set of representative optimization problems, including multimodal, unimodal, and hybrid benchmark formulations. The competitive results demonstrate the promising results of the proposed search patterns.

The main contributions of this research can be stated as follows:

1. A new physics-based optimization algorithm, namely SOA, is introduced. It uses search patterns obtained from the response of second-order systems.
2. New search patterns are proposed as an alternative to those known in the literature.
3. The statistical significance, convergence speed and exploitation-exploration ratio of SOA are evaluated against other popular metaheuristic algorithms.
4. SOA outperforms other competitor algorithms on two sets of optimization problems.

The remainder of this paper is structured as follows: A brief introduction of the second-order systems is given in Section 2; in Section 3, the most important search patterns in metaheuristic methods are discussed; in Section 4, the proposed search patterns are defined; in Section 5, the measurement of exploration-exploitation is described; in Section 6, the proposed scheme is introduced; Section 7 presents the numerical results; in Section 8, the main characteristics of the proposed approach are discussed; in Section 9, finally, the conclusions are drawn.

2. Second-Order Systems

A model whose input $R(s)$-output $C(s)$ formulation is a second-order closed-loop transfer function is known as a second-order system. One of the main elements that make a second-order model important is its ability to present very different behaviors depending on the configuration of its parameters. A generic second-order model can be formulated under the following expression [29],

$$\frac{C(s)}{R(s)} = \frac{\omega_n^2}{s^2 + 2\zeta\omega_n s + \omega_n^2},\tag{1}$$

where ζ and ω_n represent the damping ratio and ω_n the natural frequency, respectively, while s symbolizes the Laplace domain.

The dynamic behavior of a system is evaluated in terms of the temporal response obtained through a unitary step signal as input $R(s)$. The dynamic behavior is defined as the way in which the system reacts, trying to reach the value of one as time evolves. The dynamic behavior of the second-order system is described in terms of ζ and ω_n [30]. Assuming such parameters, the second-order system presents three different behaviors: Underdamped ($0 < \zeta < 1$), critically damped ($\zeta = 1$), and overdamped ($\zeta > 1$).

2.1. Underdamped Behavior ($0 < \zeta < 1$)

In this behavior, the poles (roots of the denominator) of Equation (1) are complex conjugated and located in the left-half of the s plane. Under such conditions, the system underdamped response $C_U(s)$ in the Laplace domain can be characterized as follows:

$$C_U(s) = \frac{\omega_n^2}{s(s + \zeta\omega_n)^2 + \omega_n^2(1 - \zeta^2)}.\tag{2}$$

Applying partial fraction operations and the inverse Laplace transform, it is obtained the temporal response that describe the underdamped behavior $c_U(t)$ as it is indicated in Equation (3):

$$c_U(t) = 1 - \frac{e^{-\zeta\omega_n t}}{\sqrt{1 - \zeta^2}} \sin\left(\omega_n\sqrt{1 - \zeta^2} + \tan^{-1}\left(\frac{\sqrt{1 - \zeta^2}}{\zeta}\right)\right).\tag{3}$$

If $\zeta = 0$, a special case is presented in which the temporal system response is oscillatory. The output of these behaviors is visualized in Figure 2 for the cases of $\zeta = 0$, $\zeta = 0.2$, $\zeta = 0.5$ and $\zeta = 0.707$. Under the underdamped behavior, the system response starts with high acceleration. Therefore, the response produces an overshoot that surpasses the value of one. The size of the overshoot inversely depends on the value of ζ.

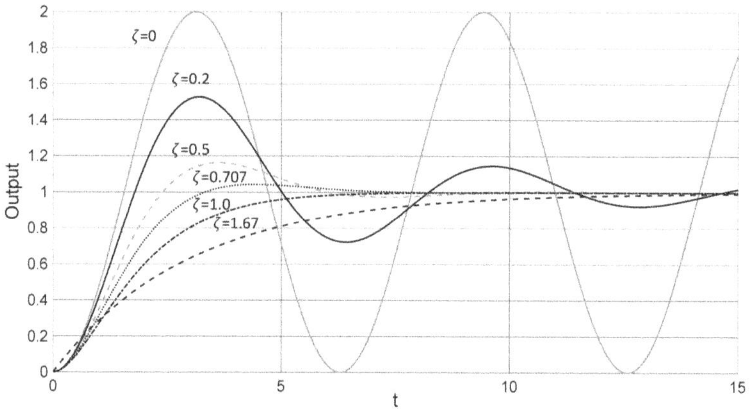

Figure 2. Temporal responses of second-order system considering its different behaviors: Underdamped $(0 < \zeta < 1)$, critically damped $(\zeta = 1)$, and overdamped $(\zeta > 1)$.

2.2. Critically Damped Behavior $(\zeta = 1)$

Under this behavior, the two poles of the transfer function of Equation (1) present a real number and maintain the same value. Therefore, the response of the critically damped behavior $C_C(s)$ in the Laplace domain can be described as follows:

$$C_C(s) = \frac{\omega_n^2}{s(s + \omega_n)^2}.\tag{4}$$

Considering the inverse Laplace transform of Equation (4), the temporal response of the critically damped behavior $c_C(t)$ is determined under the following model:

$$c_C(t) = 1 - e^{-\omega_n t}(1 + \omega_n t).\tag{5}$$

Under the critically damped behavior, the system response presents a temporal pattern similar to a first-order system. It reaches the objective value of one without experimenting with an overshoot. The output of the critically damped behavior is visualized in Figure 2.

2.3. Overdamped Behavior $(\zeta > 1)$

In the overdamped case, the two poles of a transfer function of Equation (1) have real numbers but with different values. Its response $C_O(s)$ in the Laplace domain is modeled under the following formulation:

$$C_O(s) = \frac{\omega_n^2}{s\left(s + \zeta\omega_n + \omega_n\sqrt{\zeta^2 - 1}\right)\left(s + \zeta\omega_n - \omega_n\sqrt{\zeta^2 - 1}\right)}.\tag{6}$$

After applying the inverse Laplace transform, it is obtained the temporal response of the overdamped behavior $c_O(t)$ defined as follows:

$$c_O(t) = 1 + e^{-(\zeta - \sqrt{\zeta^2 - 1})\omega_n t}.\tag{7}$$

Under the Overdamped behavior, the system slowly reacts until reaching the value of one. The deceleration of the response depends on the value of ζ. The greater the value of ζ, the slower the response will be. The output of this behavior is visualized in Figure 2 for the case of $\zeta = 1.67$.

3. Search Patterns in Metaheuristics

The generation of efficient search patterns for the correct exploration of a fitness landscape could be complicated, particularly in the presence of ruggedness and multiple local optima. Recently, several new metaheuristic schemes have been introduced in the literature. Although all these approaches consider very different phenomena as metaphors, the search patterns, used to explore the search space, are very similar. A search pattern is a set of movements produced by a rule or model in order to examine promising solutions from the search space.

Exploration and exploitation correspond to the most important characteristics of a search pattern. Exploration refers to the ability of a search pattern to examine a set of solutions spread in distinct areas of the search space. On the other hand, exploitation represents the capacity of a search pattern to improve the accuracy of the existent solutions through a local examination. The combination of both mechanisms in a search pattern is crucial for attaining success when solving a particular optimization problem.

To solve the optimization formulation, from a metaheuristic point of view, a population of $\mathbf{P}^k\left(\left\{\mathbf{x}_1^k,\ldots,\mathbf{x}_N^k\right\}\right)$ of N candidate solutions (individuals) evolve from an initial point ($k = 1$) to a *Maxgen* number of generations ($k = Maxgen$). In the population, each individual $\mathbf{x}_i^k$ ($i \in [1,\ldots,N]$) corresponds to a d-dimensional element $\left\{x_{i,1}^k,\ldots,x_{i,d}^k\right\}$, which symbolizes the decision variables involved by the optimization problem. At each generation, search patterns are applied over the individuals of the population $\mathbf{P}^k$ to produce the new population $\mathbf{P}^{k+1}$. The quality of each individual $\mathbf{x}_i^k$ is evaluated in terms of its solution regarding the objective function $J\left(\mathbf{x}_i^k\right)$ whose result represents the fitness value of $\mathbf{x}_i^k$. As the metaheuristic method evolves, the best current individual $\mathbf{b}\{b_1,\ldots,b_d\}$ is maintained since $\mathbf{b}$ represents the best available solution seen so-far.

In general, a search pattern is applied to each individual $\mathbf{x}_i^k$ using the best element $\mathbf{b}$ as a reference. Then, following a particular model, a set of movements are produced to modify the position of $\mathbf{x}_i^k$ until the location of $\mathbf{b}$ has been reached. The idea behind this mechanism is to examine solutions in the trajectory from $\mathbf{x}_i^k$ to $\mathbf{b}$ with the objective to find a better solution than the current $\mathbf{b}$. Search patterns differ in the model employed to produce the trajectories $\mathbf{x}_i^k$ from to $\mathbf{b}$.

Two of the most popular search models are attraction and spiral trajectories. The attraction model generates attraction movements from $\mathbf{x}_i^k$ to $\mathbf{b}$. The attraction model is used extensively by several metaheuristic methods such as PSO [10–12], FA [13,14], CS [15], BA [16], GSA [24], EM [27] and DE [6]. On the other hand, the spiral model produces a spiral trajectory that encircles the best element $\mathbf{b}$. The spiral model is employed by the recently published WOA and GWO schemes. Trajectories produced by the attraction, and spiral models are visualized in Figure 3a,b, respectively.

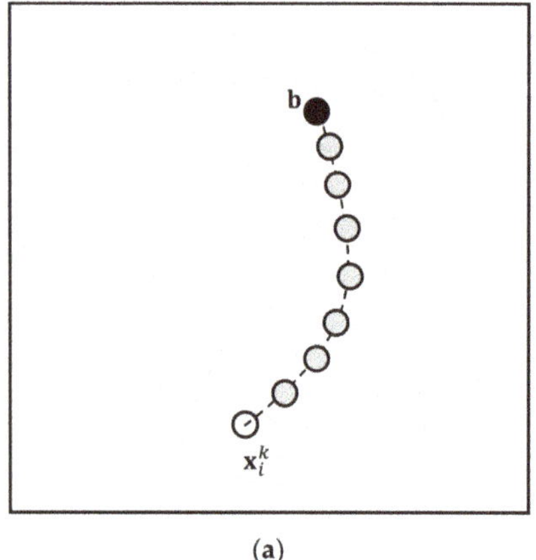
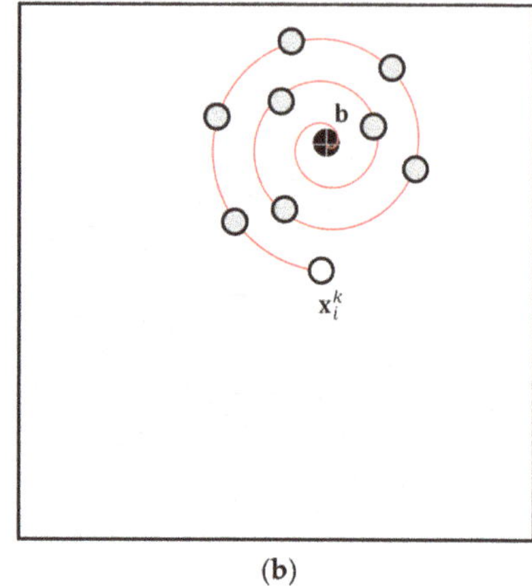

(a) **(b)**

Figure 3. Trajectories produced by, (**a**) attraction, and (**b**) spiral models.

4. Proposed Search Patterns

In this paper, a set of new search patterns are introduced to explore the search space efficiently. They emulate the response of a second-order system. The proposed set of search patterns have been integrated as a complete search strategy to obtain the global solution of complex optimization problems. Since the proposed scheme is based on the response of the second-order systems, it can be considered as a physics-based algorithm. In our approach, the temporal response of second-order system is used to generate the trajectory from the position of $\mathbf{x}_i^k = \left\{ x_{i,1}^k, \ldots, x_{i,d}^k \right\}$ to the location of $\mathbf{b} = \{b_1, \ldots, b_d\}$. With the use of such models, it is possible to produce more complex trajectories that allow a better examination of the search space. Under such conditions, we consider the three different responses of a second-order system to produce three distinct search patterns. They are the underdamped, critically damped and overdamped modeled by the expressions Equations (8)–(10), respectively:

$$x_{i,j}^k = \left(1 - \frac{e^{-\zeta \omega_n k}}{\sqrt{1 - \zeta^2}} \sin \left(\omega_n \sqrt{1 - \zeta^2} + \tan^{-1} \left(\frac{\sqrt{1 - \zeta^2}}{\zeta} \right) \right) \right) \left(b_j - x_{i,j}^k \right); \tag{8}$$

$$x_{i,j}^k = \left(1 - e^{-\omega_n k}(1 + \omega_n k) \right) \left(b_j - x_{i,j}^k \right); \tag{9}$$

$$x_{i,j}^k = \left(1 + e^{-(\zeta - \sqrt{\zeta^2 - 1})\omega_n k} \right) \left(b_j - x_{i,j}^k \right); \tag{10}$$

where i ($\in [1, N]$) corresponds to the search agent while j ($\in [1, d]$) symbolizes the decision variable or dimension. Since the behavior of each search pattern depends on the value of ζ, it is easy to combine elements to produce interesting trajectories. Figure 4 presents some examples of trajectories produced by using different values for ζ. In the Figure, it is assumed a two-dimensional case ($d = 2$) where the initial position of the search agent $\mathbf{x}_i^k$ is $(0.5, 0.5)$ and the final location or the best location $(1, 1)$. Figure 4a presents the case of $x_{i,1}^k \leftarrow \zeta = 0$ and $x_{i,2}^k \leftarrow \zeta = 1$. Figure 4b presents the case of $x_{i,1}^k \leftarrow \zeta = 0.1$ and $x_{i,2}^k \leftarrow \zeta = 0.5$. Figure 4c presents the case of $x_{i,1}^k \leftarrow \zeta = 1$ and $x_{i,2}^k \leftarrow \zeta = 1.67$. Finally, Figure 4d presents the case of $x_{i,1}^k \leftarrow \zeta = 0.5$ and $x_{i,2}^k \leftarrow \zeta = 1$. From the figures, it is clear that the second-order responses allow producing several complex trajectories, which include most of the other search patterns known in the literature. In all cases (a)–(d), the value of ω_n has been set to 1.

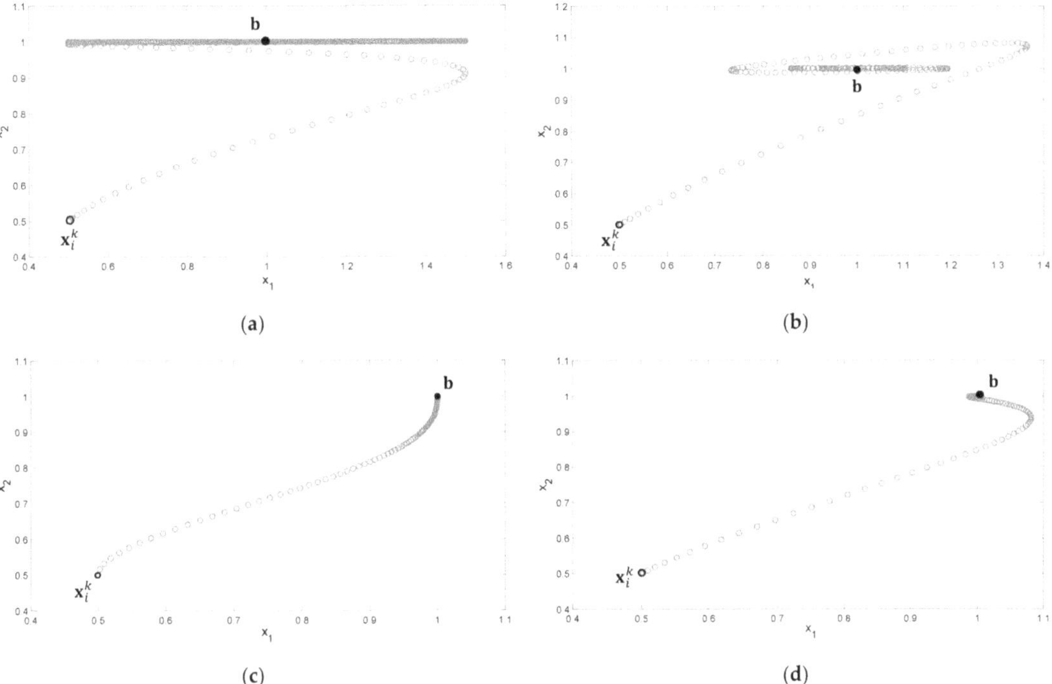

Figure 4. Some examples of trajectories produced by using different values for ζ. (a) $x_{i,1}^k \leftarrow \zeta = 0$ and $x_{i,2}^k \leftarrow \zeta = 1$, (b) $x_{i,1}^k \leftarrow \zeta = 0.1$ and $x_{i,2}^k \leftarrow \zeta = 0.5$, (c) $x_{i,1}^k \leftarrow \zeta = 1$ and $x_{i,2}^k \leftarrow \zeta = 1.67$ and (d) $x_{i,1}^k \leftarrow \zeta = 0.5$ and $x_{i,2}^k \leftarrow \zeta = 1$.

5. Balance of Exploration and Exploitation

Metaheuristic methods employ a set of search agents to examine the search space with the objective to identify a satisfactory solution for an optimization formulation. In metaheuristic schemes, search agents that present the best fitness values tend to regulate the search process, producing an attraction towards them. Under such conditions, as the optimization process evolves, the distance among individuals diminishes while the effect of exploitation is highlighted. On the other hand, when the distance among individuals increases, the characteristics of the exploration process are more evident.

To compute the relative distance among individuals (increase and decrease), a diversity indicator known as the dimension-wise diversity index [31] is used. Under this approach, the diversity is formulated as follows,

$$Div_j = \frac{1}{N} \sum_{i=1}^{N} \left| median\left(x^j\right) - x_{i,j} \right| Div = \frac{1}{d} \sum_{j=1}^{d} Div_j \tag{11}$$

where $median\left(x^j\right)$ symbolizes the median of dimension j of all search agents. $x_{i,j}$ represents the variable decision j of the individual i. N is the number of individuals in the population $\mathbf{P}^k$ while d corresponds to the number of dimensions of the optimization formulation. The diversity Div_j (of the j-th dimension) evaluates the relative distance between the variable j of each individual and its median value. The complete diversity Div (of the entire population) corresponds to the averaged diversity in each dimension. Both elements Div_j and Div are calculated in every iteration.

Having evaluated the diversity values, the level of exploration and exploitation can be computed as the percentage of the time that a search strategy invests exploring or exploiting in terms of its diversity values. These percentages are calculated in each iteration by means of the following models,

$$XPL\% = \left(\frac{Div}{Div_{max}} \right) \times 100 \qquad XPT\% = \left(\frac{|Div - Div_{max}|}{Div_{max}} \right) \times 100 \tag{12}$$

where Div_{max} symbolizes the maximum diversity value obtained during the optimization process. The percentage of exploration $XPL\%$ corresponds to the size of exploration as the rate between Div and Div_{max}. On the other hand, the percentage of exploitation $XPT\%$ symbolizes the level of exploitation. $XPT\%$ is computed as the complemental percentage to $XPL\%$ since the difference between Div_{max} and Div is generated because of the concentration of individuals.

6. Proposed Metaheuristic Algorithm

The set of search patterns based on the second-order systems have been integrated as a complete search strategy to obtain the global solution of complex optimization problems. In this section, the complete metaheuristic method, called Second-Order Algorithm (SOA), is completely described.

The scheme considers four different stages: (A) Initialization, (B) trajectory generation, (C) reset of bad elements, and (D) avoid premature convergence mechanism. The steps (B)–(D) are sequentially executed until a stop criterion has been reached. Figure 5 shows the flowchart of the complete metaheuristic method.

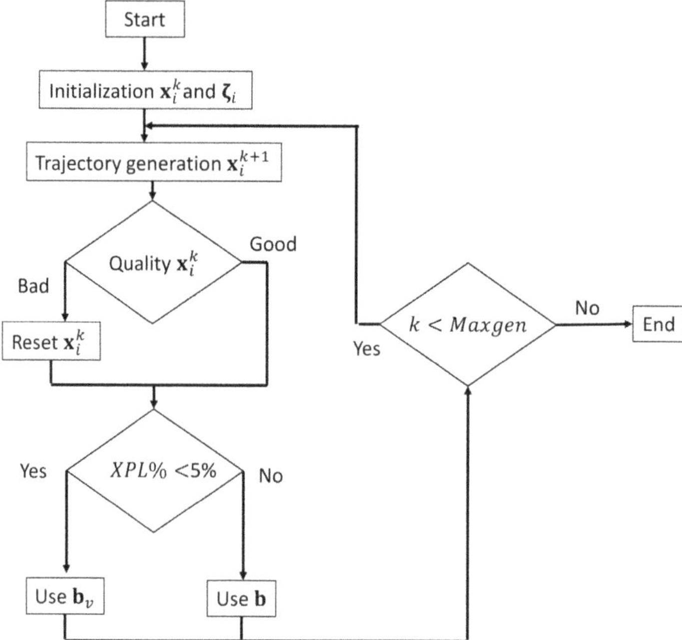

Figure 5. Flowchart of the proposed metaheuristic method based on the response of second-order systems.

6.1. Initialization

In the first iteration $k = 0$, a population $\mathbf{P}^0$ of N agents $\{\mathbf{x}_1^0, \ldots, \mathbf{x}_N^0\}$ is randomly produced considering to the following equation,

$$x_{i,j}^0 = rand \cdot \left(b_j^{high} - b_j^{low} \right) + b_j^{low} \qquad i = 1, 2, \ldots, N; \ j = 1, \ldots, d \tag{13}$$

where b_j^{high} and b_j^{low} are the limits of the j decision variable and *rand* is a uniformly distributed random number between [0,1].

To each individual $\mathbf{x}_i$ from the population, it is assigned a vector $\zeta_i = \left\{ \zeta_{i,1}, \ldots, \zeta_{i,d} \right\}$ whose elements $\zeta_{i,j}$ determine the trajectory behavior of each j-th dimension. Initially, each element $\zeta_{i,j}$ is set to a random value between [0,2]. Under this interval, all the second-order behavior are possible: Underdamped ($0 < \zeta < 1$), critically damped ($\zeta = 1$), and overdamped ($\zeta > 1$).

6.2. Trajectory Generation

Once the population has been initialized, it is obtained the best element of the population $\mathbf{b}$. Then, the new position $\mathbf{x}_i^{k+1}$ of each agent $\mathbf{x}_i^k$ is computed as a trajectory generated by a second-order system. Once all new positions in the population $\mathbf{P}^k$ are determined, it is also defined the best element $\mathbf{b}$.

6.3. Reset of Bad Elements

To each agent $\mathbf{x}_i^k$ is allowed to move in its own trajectory for ten iterations. After ten iterations, if the search agent $\mathbf{x}_i^k$ maintains the worst performance in terms of the fitness function, it is reinitialized in both position and in its vector ζ_i. Under such conditions, the search agent will be in another position and with the ability to perform another kind of trajectory behavior.

6.4. Avoid Premature Convergence Mechanism

If the percentage of exploration $XPL\%$ is less than 5%, the best value $\mathbf{b}$ is replaced by the best virtual value $\mathbf{b}_v$. The element $\mathbf{b}_v$ is computed as the averaged value of the best five individuals of the population. The idea behind this mechanism is to identify a new position to generate different trajectories that avoid that the search process gets trapped in a local optimum.

7. Experimental Results

To evaluate the results of the proposed SOA algorithm, a set of experiments has been conducted. Such results have been compared to those produced by the Artificial Bee Colony (ABC) [9], the Covariance matrix adaptation evolution strategy (CMAES) [4], the Crow Search Algorithm (CSA) [8], the Differential Evolution (DE) [6], the Moth-flame optimization algorithm (MFO) [18] and the Particle Swarm Optimization (PSO) [10], which are considered the most popular metaheuristic schemes in many optimization studies [32].

For the comparison, all methods have been set according to their reported guidelines. Such configurations are described as follows:

- ABC: Onlooker Bees = 50, acceleration coefficient = 1 [9].
- DE: crossover probability = 0.2, Betha = 1 [6].
- CMAES: Lambda = 50, father number = 25, sigma = 60, csigma = 0.32586, dsigma = 1.32586 [4].
- CSA: Flock = 50, awareness probability = 0.1, flight length = 2 [8].
- MFO: search agents = 50, "a" linearly decreases from 2 to 0 [18].
- SOA: the experimental results give the best algorithm performance with the next parameter set par1 = 0.7, par2 = 0.3 and par3 = 0.05.

In our analysis, the population size N has been set to 50 search agents. The maximum iteration number ($Maxgen$) for all functions has been set to 1000. This stop criterion has been decided to keep compatibility with similar works published in the literature [33,34]. To evaluate the results, three different indicators are considered: The Average Best-so-far (**AB**) solution, the Median Best-so-far (**MB**) solution and the Standard Deviation (**SD**) of the best-so-far solutions. In the analysis, each optimization problem is solved using every algorithm 30 times. From this operation, 30 results are produced. From all these values, the mean value of all best-found solutions represents the Average Best-so-far (**AB**) solution.

Likewise, the median of all 30 results is computed to generate **MB** and the standard deviation of the 30 data is estimated to obtain **SD** of the best-so-far solutions. Indicators **AB** and **MB** correspond to the accuracy of the solutions, while **SD** their dispersion, and thus, the robustness of the algorithm.

The experimental section is divided into five sub-sections. In the first Section 7.1, the performance of SOA is evaluated with regard to multimodal functions. In the second Section 7.2, the results of the OTSA method in comparison with other similar approaches are analyzed in terms of unimodal functions. In the third Section 7.3, a comparative study among the algorithms examining hybrid functions is accomplished. In the fourth Section 7.4, the ability of all algorithms to converge is analyzed. Finally, in the fifth Section 7.5, the performance of the SOA method to solve the CEC 2017 set of functions is also analyzed.

7.1. Multimodal Functions

In this sub-section, the SOA approach is evaluated considering 12 unimodal functions $(f_1(\mathbf{x})-f_{12}(\mathbf{x}))$ reported in Table 1 from Appendix A. Multimodal functions present optimization surfaces that involve multiple local optima. For this reason, these function presents more complications in their solution. In this analysis, the performance of the SOA method is examined in comparison with ABC, CMAES, CSA, DE, MFO and PSO in terms of the multimodal functions. Multimodal objective functions correspond to functions from $f_1(\mathbf{x})$ to $f_{12}(\mathbf{x})$ in Table 1 from the Appendix A, where the set of local minima augments as the dimension of the function also increases. Therefore, the study exhibits the capacity of each metaheuristic scheme to identify the global optimum when the function contains several local optima. In the experiments, it is assumed objective functions operating in 30 dimensions ($n = 30$). The averaged best (AB) results considering 30 independent executions are exhibit in Table 1. It also reports the median values (MD) and the standard deviations (SD).

Table 1. Minimization results of multimodal benchmark functions.

		ABC	DE	CMAES	CSA	PSO	MFO	SOA
	AB	8.9132622	0.7932535	2.8976×10^{-19}	55.918504	0.2012803	27.983271	0.1119774
$f_1(\mathbf{x})$	MD	8.4392750	0.7993166	2.4779×10^{-19}	57.038263	4.1459×10^{-23}	25.365199	1.0714×10^{-10}
	SD	2.6748059	0.1378538	1.5343×10^{-19}	6.2032578	1.1022968	12.283254	0.2272439
	AB	2	2	2	1,897,783.3	27.4	2	2
$f_2(\mathbf{x})$	MD	2	2	2	35,691.155	2	2	2
	SD	9.9512×10^{-12}	0	0	9,636,632.1	113.43495	0	0
	AB	2	2	2	3,620,834.4	34.723128	2	2
$f_3(\mathbf{x})$	MD	2	2	2	676,981.46	9	2	2
	SD	2.3308×10^{-11}	0	0	7,265,352.7	113.36672	0	0
	AB	0.1371551	0.002	1.7942×10^{-6}	0.0862919	2.2285×10^{-8}	5.5194×10^{-10}	1.164×10^{-11}
$f_4(\mathbf{x})$	MD	0.1349076	0.01	0	0.0892256	0	0	7.7118×10^{-12}
	SD	0.0399861	0.123	5.4833×10^{-6}	0.0213332	1.2206×10^{-7}	3.0231×10^{-9}	1.0995×10^{-11}
	AB	13,781,291	1,331,987.7	22,307.195	44,274,761	82.539625	85.756149	71.964984
$f_5(\mathbf{x})$	MD	13,876,263	1,365,502.1	72.377516	46,153,728	81.698488	85.665615	72.362277
	SD	3,237,147.7	306,385.34	50,923.637	10,118,180	7.1916726	3.2857088	0.9929591
	AB	1.152×10^{85}	5.850×10^{81}	1.812×10^{83}	6.429×10^{83}	1.397×10^{81}	3.0883×10^{81}	1.0051×10^{81}
$f_6(\mathbf{x})$	MD	4.622×10^{84}	3.405×10^{81}	7.928×10^{82}	2.939×10^{83}	5.977×10^{80}	7.9607×10^{80}	4.901×10^{80}
	SD	1.685×10^{85}	7.33×10^{81}	3.047×10^{83}	7.551×10^{83}	2.197×10^{81}	5.6651×10^{81}	1.9457×10^{81}
	AB	30.033333	30	30	58.766666	30	33.633333	30
$f_7(\mathbf{x})$	MD	30	30	30	59	30	30	30
	SD	0.18257419	0	0	1.77498583	0	5.4550409	0
	AB	9.2	8.0666666	1.0666666	19,543.266	0.0333333	2000.0333	0
$f_8(\mathbf{x})$	MD	9	8	0	19,797	0	0	0
	SD	2.5784250	1.9464084	3.1941037	2077.7459	0.1825741	4842.3277	0
	AB	−745.05202	−1125.4815	−1127.8626	−725.09353	−1071.7869	−1031.2617	−1146.3478
$f_9(\mathbf{x})$	MD	−743.4462	−1174.9722	−1132.5748	−719.10777	−1068.9596	−1033.6178	−1145.2467
	SD	25.137593	78.706	25.809999	25.815652	34.055847	34.244188	10.928362

Table 1. *Cont.*

		ABC	DE	CMAES	CSA	PSO	MFO	SOA
$f_{10}(\mathbf{x})$	AB	110,282.54	665,278.86	-4930	1,170,939.0	45,556.260	222,833.73	-501.79356
	MD	96,461.061	673,449.49	-4930	1,126,234.2	5076.8152	71,582.051	-332.82466
	SD	44,933.417	129,147.27	3.7318×10^{-9}	159,175.64	75,990.880	305,159.37	663.92006
$f_{11}(\mathbf{x})$	AB	-18.26109	-26.056561	-29.6576	-16.504756	-28.367666	-28.863589	-30
	MD	-18.131984	-26.092349	-29.9286	-16.183447	-28.14029	-29.070145	-30
	SD	1.6366873	0.5428906	0.466	1.16917142	1.5408536	1.2837415	0
$f_{12}(\mathbf{x})$	AB	1502.3129	369.60375	786.36819	519.17242	196.95838	261.52332	11.905761
	MD	1457.6865	368.82916	778.72369	465.93238	213.00730	252.76747	0.3841574
	SD	420.70611	35.389136	215.93893	228.69860	86.542862	106.52353	29.544101

According to Table 1, the proposed SOA scheme obtain a better performance than ABC, CMAES, CSA, DE, MFO and PSO in functions $f_1(\mathbf{x})$, $f_4(\mathbf{x})$, $f_5(\mathbf{x})$, $f_6(\mathbf{x})$, $f_8(\mathbf{x})$, $f_9(\mathbf{x})$, $f_{10}(\mathbf{x})$, $f_{11}(\mathbf{x})$ and $f_{12}(\mathbf{x})$. Nevertheless, the results of SOA exhibit similar as the obtained by DE, CMAES and MFO in functions $f_2(\mathbf{x})$, $f_3(\mathbf{x})$ and $f_7(\mathbf{x})$.

To statistically validate the conclusions from Table 1, a non-parametric study is considered. In this test, the Wilcoxon rank-sum analysis [35] is adopted with the objective to validate the performance results. This statistical test evaluates if exists a significant difference when two methods are compared. For this reason, the analysis is performed considering a pairwise comparison such as SOA versus ABC, SOA versus CMAES, SOA versus CSA, SOA versus DE, SOA versus MFO and SOA versus PSO. In the Wilcoxon analysis, a null hypothesis (H_0) was adopted that showed that there is no significant difference in the results. On the other hand, it is assumed as an alternative hypothesis (H_1) that the result has a similar structure. For the Wilcoxon analysis, it is assumed a significance value of 0.05 considering 30 independent execution for each test function. Table 2 shows the p-values assuming the results of Table 2 (where $n = 30$) produced by the Wilcoxon study. For faster visualization, in the Table, we use the following symbols ▲ ▼, and ▶. The symbol ▲ refers that the SOA algorithm produces significantly better solutions than its competitor. ▼ symbolizes that SOA obtains worse results than its counterpart. Finally, the symbol ▶ denotes that both compared methods produce similar solutions. A close inspection of Table 2 demonstrates that for functions f_1, f_4, f_5, f_6, f_8, f_9, f_{10}, f_{11} and f_{12} the proposed SOA scheme obtain better solutions than the other methods. On the other hand, for functions f_2, f_2 and f_7, it is clear that the groups SOA versus ABC, SOA versus CMAES, SOA versus DE and SOA versus MFO and EA-HC versus SCA present similar solutions.

Table 2. Wilcoxon analysis for multimodal benchmark functions.

Function	SOA vs. ABC	SOA vs. CMAES	SOA vs. CSA	SOA vs. DE	SOA vs. MFO	SOA vs. PSO
$f_1(x)$	7.13×10^{-9} ▲	2.61×10^{-8} ▲	2.40×10^{-11} ▲	9.77×10^{-7} ▲	3.97×10^{-11} ▲	6.87×10^{-8} ▲
$f_2(x)$	1.21×10^{-12} ▶	1 ▶	1.21×10^{-12} ▲	1 ▶	1 ▶	4.13×10^{-9} ▲
$f_3(x)$	1.21×10^{-12} ▶	1 ▶	1.21×10^{-12} ▲	1 ▶	1 ▶	8.33×10^{-7} ▲
$f_4(x)$	6.48×10^{-12} ▲	2.43×10^{-8} ▲	6.48×10^{-12} ▲	1.10×10^{-7} ▲	1.18×10^{-7} ▲	5.79×10^{-8} ▲
$f_5(x)$	3.02×10^{-11} ▲	2.18×10^{-6} ▲	3.02×10^{-11} ▲	3.02×10^{-11} ▲	8.30×10^{-1} ▲	9.94×10^{-8} ▲
$f_6(x)$	3.69×10^{-11} ▲	1.29×10^{-9} ▲	2.87×10^{-10} ▲	6.73×10^{-8} ▲	1.75×10^{-5} ▲	9.52×10^{-4} ▲
$f_7(x)$	3.34×10^{-1} ▶	1 ▶	1.57×10^{-12} ▲	3.34×10^{-1} ▶	2.23×10^{-5} ▲	3.34×10^{-7} ▲
$f_8(x)$	3.96×10^{-6} ▲	1.10×10^{-7} ▲	7.87×10^{-12} ▲	3.28×10^{-6} ▲	2.45×10^{-1} ▲	5.58×10^{-7} ▲
$f_9(x)$	2.97×10^{-11} ▲	5.75×10^{-8} ▲	2.97×10^{-11} ▲	7.72×10^{-8} ▲	4.20×10^{-4} ▲	1.83×10^{-5} ▲
$f_{10}(x)$	3.02×10^{-11} ▲	2.85×10^{-11} ▲	3.02×10^{-11} ▲	3.02×10^{-11} ▲	2.57×10^{-7} ▲	1.86×10^{-3} ▲
$f_{11}(x)$	2.80×10^{-11} ▲	1.12×10^{-07} ▶	2.80×10^{-11} ▲	3.00×10^{-11} ▲	8.88×10^{-1} ▶	8.86×10^{-6} ▲
$f_{12}(x)$	3.02×10^{-11} ▲	3.02×10^{-11} ▲	3.02×10^{-11} ▲	3.02×10^{-11} ▲	1.78×10^{-10} ▲	9.76×10^{-10} ▲

7.2. Unimodal Functions

In this subsection, the performance of SOA is compared with ABC, DE, DE, CMAES CSA and MFO, considering four unimodal functions with only one optimum. Such functions are represented by functions from $f_{13}(\mathbf{x})$ to $f_{16}(\mathbf{x})$ in Table 1. In the test, all functions are considered in 30 dimensions ($d = 30$). The experimental results, obtained from 30 independent executions, are presented in Table 3. They report the results in terms of **AB**, **MB** and **SD** obtained in the executions. According to Table 3, the SOA approach provides better performance than ABC, DE, DE, CMAES CSA and MFO for all functions. In general, this study demonstrates big differences in performance among the metaheuristic scheme, which is directly related to a better trade-off between exploration and exploitation produced by the trajectories of the SOA scheme. Considering the information from Table 3, Table 4 reports the results of the Wilcoxon analysis. An inspection of the p-values from Table 4, it is clear that the proposed SOA method presents a superior performance than each metaheuristic algorithm considered in the experimental study.

Table 3. Minimization results of unimodal benchmark functions.

		ABC	DE	CMAES	CSA	PSO	MFO	SOA
$f_{13}(\mathbf{x})$	AB	25.648891	23.234006	0.0382940	117,864.48	2433.8148	16,893.538	4.416×10^{-16}
	MD	26.054364	22.218747	1.503×10^{-23}	120,885.37	5.9145×10^{-10}	10,737.418	4.1862×10^{-16}
	SD	8.3305663	5.5158397	0.1428856	14,742.855	4322.0341	19,501.276	2.4373×10^{-16}
$f_{14}(\mathbf{x})$	AB	0.0136600	0.0145526	1.2398×10^{-5}	51.476735	4.0467×10^{-13}	7.8643202	1.5×10^{-20}
	MD	0.0144199	0.0144921	1.3237×10^{-20}	51.798031	7.5065×10^{-14}	2.8398×10^{-7}	1.3059×10^{-20}
	SD	0.0041468	0.0037530	3.908×10^{-5}	5.9378289	7.8227×10^{-13}	14.024260	9.6811×10^{-21}
$f_{15}(\mathbf{x})$	AB	0.5442777	0.569612	19.789197	2466.2017	20	403.33365	1.2053×10^{-18}
	MD	0.4952416	0.5733913	0.2257190	2478.7246	8.9126×10^{-12}	200.00000	1.025×10^{-18}
	SD	0.1986349	0.1376955	39.959487	344.56578	66.436383	520.92974	6.4946×10^{-19}
$f_{16}(\mathbf{x})$	AB	0.0006647	1.8659×10^{-10}	6.9743×10^{-10}	0.0068342	5.6588×10^{-24}	9.5555×10^{-19}	0
	MD	0.0004306	1.2946×10^{-10}	7.1112×10^{-10}	0.0066858	3.6687×10^{-29}	2.8339×10^{-22}	0
	SD	0.0006187	1.8814×10^{-10}	4.3714×10^{-10}	0.0032037	3.0973×10^{-23}	3.3608×10^{-18}	0

Table 4. Wilcoxon analysis for unimodal benchmark functions.

Function	SOA vs. ABC	SOA vs. CMAES	SOA vs. CSA	SOA vs. DE	SOA vs. MFO	SOA vs. PSO
$f_{13}(\mathbf{x})$	2.80×10^{-11} ▲	3.86×10^{-1} ▲	2.80×10^{-11} ▲	2.80×10^{-11} ▲	1.75×10^{-9} ▲	1.22×10^{-4} ▲
$f_{14}(\mathbf{x})$	5.51×10^{-9} ▲	2.11×10^{-1} ▲	2.72×10^{-11} ▲	3.22×10^{-9} ▲	1.07×10^{-4} ▲	1.74×10^{-2} ▼
$f_{15}(\mathbf{x})$	1.58×10^{-1} ▲	3.02×10^{-11} ▲	3.02×10^{-11} ▲	1.81×10^{-1} ▲	5.26×10^{-4} ▲	3.11×10^{-1} ▲
$f_{16}(\mathbf{x})$	1.21×10^{-12} ▲	1.21×10^{-12} ▲	1.21×10^{-12} ▲	1.21×10^{-12} ▲	1.21×10^{-12} ▲	1.21×10^{-12} ▲

7.3. Hybrid Functions

In this study, hybrid functions are used to evaluate the optimization solutions of the SOA scheme. Hybrid functions refer to multimodal optimization problems produced by the combination of several multimodal functions. These functions correspond to the formulations from $f_{17}(\mathbf{x})$ to $f_{20}(\mathbf{x})$, which are shown in Table 1 in Appendix A. In the experiments, the performance of our proposed SOA approach has been compared with other metaheuristic schemes.

The simulation results are reported in Table 5. It exhibits the performance of each algorithm in terms of **AB**, **MB** and **SD**. From Table 5, it can be observed that the SOA method presents a superior performance than the other techniques in all functions. Table 6 reports the results of the Wilcoxon analysis assuming the index of the Average Best-so-far (**AB**) values of Table 5. Since all elements present the symbol ▲, they validate that the proposed SOA method produces better results than the other methods. The remarkable performance of the proposed SOA scheme for hybrid functions is attributed to a better balance between exploration and exploitation of its operators provoked by the properties of the second system trajectories. This denotes that the SOA approach generates an appropriate number of promising search agents that allow an adequate exploration of the search space. On the other hand, a balanced number of candidate solutions is also produced that make it possible to improve the quality of the already-detected solutions, in terms of the objective function.

Table 5. Minimization results of hybrid benchmark functions.

		ABC	DE	CMAES	CSA	PSO	MFO	SOA
$f_{17}(x)$	AB	396.75458	7.7178366	3.1526×10^{-9}	20,330.245	334.63082	23,758.790	0.8147792
	MD	210.31173	7.7772734	2.7959×10^{-9}	19,814.613	6.9181×10^{-7}	20,077.849	4.3564×10^{-11}
	SD	497.02847	1.3190384	1.1535×10^{-9}	2074.4742	1832.8485	18,166.603	2.5336865
$f_{18}(x)$	AB	212.40266	75.917033	105.99474	731.38151	65.728942	161.36081	30.785661
	MD	212.09269	75.575979	31.783896	741.08639	65.904649	116.86683	28.998449
	SD	25.805145	10.604761	84.809720	68.759154	14.502755	107.40820	3.5251022
$f_{19}(x)$	AB	221,724.73	1264.0862	57.570417	70,494,770	87.951413	80.887783	31.999808
	MD	200,128.19	1278.1579	32.661016	65,110,859	84.267959	78.040599	31.999808
	SD	114,341.27	268.82145	54.602147	23,890,494	25.256576	23.169503	6.4582×10^{-10}
$f_{20}(x)$	AB	319.57592	49.741282	97.542580	867.75158	122.92508	802.21444	30.307556
	MD	299.50312	49.708103	29.002196	879.05543	65.254961	685.34608	29
	SD	64.709010	4.4720540	93.393879	103.61875	127.74698	500.43174	4.0252793

Table 6. Wilcoxon analysis for hybrid benchmark functions.

Function	SOA vs. ABC	SOA vs. CMAES	SOA vs. CSA	SOA vs. DE	SOA vs. MFO	SOA vs. PSO
$f_{17}(x)$	4.35×10^{-11}▲	6.63×10^{-5}▲	2.92×10^{-11}▲	8.16×10^{-8}▲	5.40×10^{-10}▲	6.55×10^{-2}▲
$f_{18}(x)$	1.16×10^{-7}▲	7.28×10^{-4}▲	3.02×10^{-11}▲	6.63×10^{-7}▲	2.01×10^{-1}▲	5.20×10^{-6}▲
$f_{19}(x)$	3.02×10^{-11}▲	3.02×10^{-11}▲	3.02×10^{-11}▲	3.02×10^{-11}▲	1.17×10^{-4}▲	4.94×10^{-5}▲
$f_{20}(x)$	4.91×10^{-11}▲	2.71×10^{-5}▲	2.98×10^{-11}▲	6.73×10^{-5}▲	5.43×10^{-11}▲	8.11×10^{-5}▲

7.4. Convergence Analysis

The evaluation of accuracy in the final solution cannot completely assess the abilities of an optimization algorithm. On the other hand, the convergence of a metaheuristic scheme represents an important property to assess its performance. This analysis determined the velocity, which determined metaheuristic scheme reaches the optimum solution. In this subsection, a convergence study has been carried out. In the comparisons, for the sake of space, the performance of the best four metaheuristic schemes is considered adopting a representative set of six functions (two multimodal, two unimodal and two hybrids), operated in 30 dimensions. To generate the convergence graphs, the raw simulation data produced in the different experiments was processed. Since each simulation is executed 30 times for each metaheuristic method, the convergence data of the execution corresponds to the median result. Figures 6–8 show the convergence graphs for the four best-performing metaheuristic methods. A close inspection of Figure 6 demonstrates that the proposed SOA scheme presents a better convergence than the other algorithms for all functions.

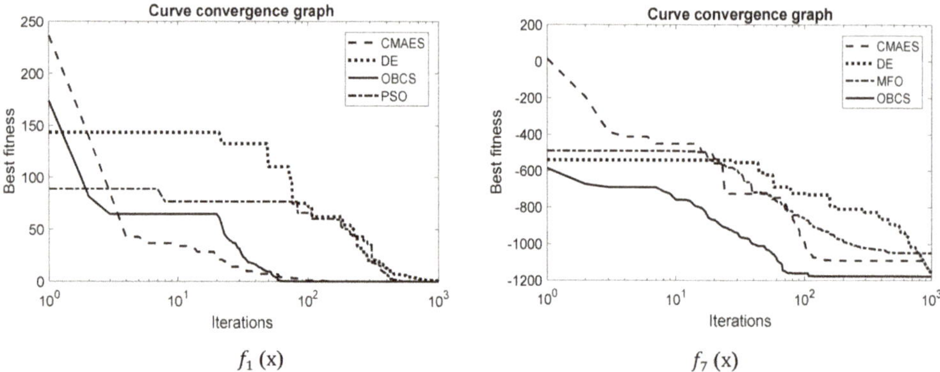

Figure 6. Convergence graphs in two representative multimodal-functions.

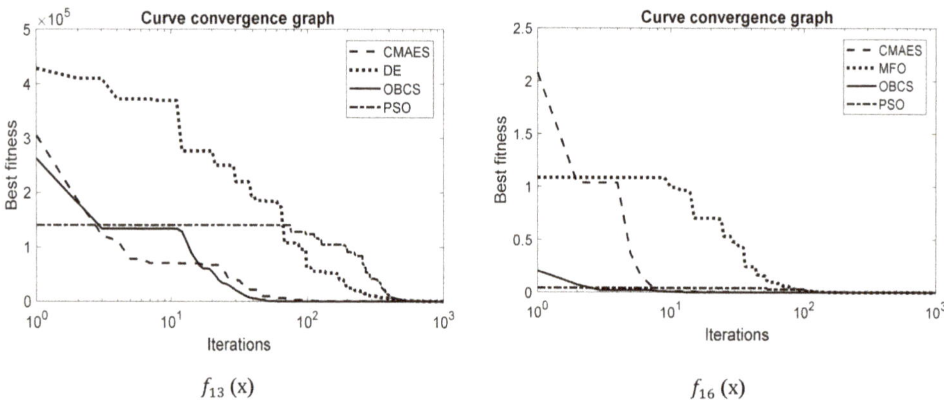

Figure 7. Convergence graphs in two representative unimodal functions.

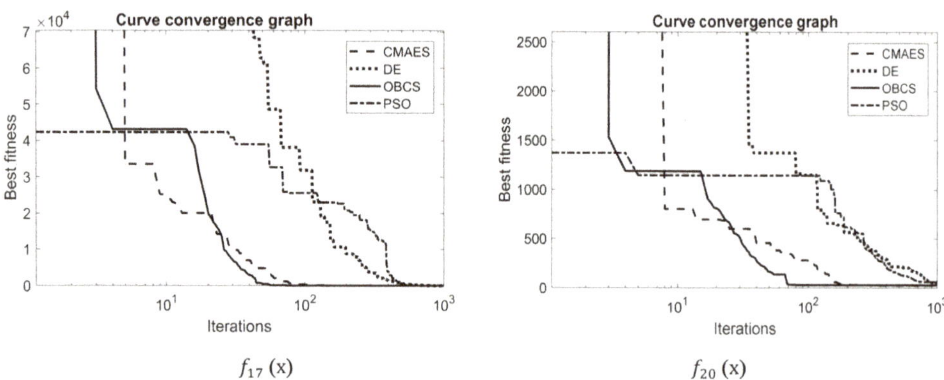

Figure 8. Convergence graphs in two representative hybrid functions.

7.5. Performance Evaluation with CEC 2017

In this sub-section, the performance of the SOA method to solve the CEC 2017 set of functions is also analyzed. The set of functions from the CEC2017 [36] represents one the most elaborated platform for benchmarking and comparing search strategies for numerical optimization. The CEC2017 benchmarks correspond to a test environment of 30 different functions with distinct features. They will be identified from $F_1(\mathbf{x})$ to $F_{30}(\mathbf{x})$. Most of these functions are similar to those exhibited in Appendix A, but with different translations and/or rotations effects. The average obtained results, corresponding to 30 independent executions, are re-registered in Table 7. The results are reported in terms of the performance indexes: Average Best fitness (AB), Median Best fitness (MB), and the standard deviation of the best finesses (SD).

Table 7. Optimization results from benchmark functions of CEC2017.

		ABC	CMAES	CSA	DE	MFO	PSO	SOA
$F_1(x)$	AB	31,543,508.93	16,816,680,165	62,350,211,795	92,576,614.71	9,869,100,268	5,754,474,631	389,770,985.4
	MD	31,272,497.07	10,317,001,637	62,180,271,044	90,540,094.66	8,120,831,786	5,716,932,117	348,397,609.3
	SD	11,711,445.55	18,287,769,833	6,701,490,748	19,876,885.3	6,276,129,768	3,760,493,838	175,590,731.3
$F_2(x)$	AB	4.9441×10^{32}	1.9584×10^{42}	9.584×10^{43}	4.6409×10^{32}	1.32094×10^{37}	1.8354×10^{43}	2.4753×10^{19}
	MD	8.92×10^{31}	7.2683×10^{41}	5.1378×10^{42}	9.165×10^{31}	5.8224×10^{31}	1.5174×10^{31}	4.8076×10^{17}
	SD	8.1141×10^{32}	3.0961×10^{42}	2.7242×10^{44}	9.8904×10^{32}	5.35797×10^{37}	1.0053×10^{44}	1.2579×10^{20}
$F_3(x)$	AB	143,585.796	206,258.738	105,227.778	184,983.575	141,570.4456	79,667.257	50,153.7766
	MD	143,614.585	201,904.141	104,066.094	187,839.624	132,229.1115	69,068.2299	47,032.1654
	SD	17,888.2525	52,000.2984	14,953.9048	27,184.1578	57,378.84773	39,077.8042	20,560.6136
$F_4(x)$	AB	558.75833	3855.6622	16,385.5764	558.09385	1026.441012	937.069787	547.952065
	MD	562.142677	3566.57161	16,520.3929	556.117066	856.5465617	890.326594	544.342148
	SD	20.2914617	1429.47447	3040.62335	23.4459834	629.655489	361.015317	20.9881356
$F_5(x)$	AB	730.381143	825.655628	951.170114	721.540918	692.7089629	629.566432	631.994737
	MD	732.333214	847.872393	951.759754	720.602523	694.7441044	634.793518	630.746916
	SD	12.7345432	66.2497783	22.7361791	9.29337642	42.18671231	29.110803	26.6185453
$F_6(x)$	AB	603.83984	669.750895	691.817816	604.901193	632.8602211	612.05324	609.329809
	MD	603.786524	668.724864	691.322797	604.969262	632.4713596	611.116445	608.748467
	SD	0.60630828	9.3375723	6.33324596	0.550883	8.875276889	5.83130625	2.23326029
$F_7(x)$	AB	977.074304	889.858004	1868.59237	983.896545	1085.414574	850.625181	933.44086
	MD	977.307172	899.416046	1847.81046	988.619865	1067.152488	837.191947	939.402368
	SD	13.674818	39.0420544	125.687304	17.0217457	139.9764841	43.9666771	25.6723087
$F_8(x)$	AB	1033.74747	1047.6713	1181.82055	1023.57155	991.3574401	914.911462	920.212894
	MD	1035.62163	1026.79246	1181.41759	1023.90313	992.7702619	916.328125	921.260403
	SD	13.5776161	86.6169315	29.0041369	12.1336359	43.34646427	24.9852693	21.5473475
$F_9(x)$	AB	1926.96476	900	15,096.7081	6434.26285	6487.6152	2436.07714	3251.16746
	MD	1839.55339	900	15,358.9047	6327.72188	5976.801998	2313.97194	2702.63581
	SD	333.208904	0	1637.81695	887.500363	2250.718589	920.338184	1462.45439
$F_{10}(x)$	AB	8559.09425	8026.51416	8695.92279	7235.23129	5260.902021	5136.84827	4524.56624
	MD	8598.36183	7995.60926	8710.13364	7246.56367	5291.679788	4941.91541	4512.99611
	SD	327.683849	246.346496	312.712801	234.084094	711.3287215	845.232077	357.978315
$F_{11}(x)$	AB	1594.97778	19,382.0122	7888.11046	1813.68615	4011.754904	1465.88225	1248.42954
	MD	1594.04404	18,978.0433	7566.7015	1774.91924	2427.207284	1465.34072	1247.83322
	SD	90.044155	9762.53738	2030.84932	246.845029	3525.449844	125.920596	32.5337336
$F_{12}(x)$	AB	22,035,530.6	4,281,754,760	8,661,838,227	92,723,516.6	91,292,958.57	354,557,569	4,401,381.17
	MD	20,991,516.3	4,358,279,185	8,610,836,607	93,208,101.7	23,663,273.6	246,467,712	3,700,514.39
	SD	7,163,344.91	1,489,535,737	2,020,262,530	18,696,916.2	134,545,916.4	411,606,206	3,405,855.44
$F_{13}(x)$	AB	19,266.9698	3,652,661,382	6,874,450,894	3,979,238.98	38,881,437.05	111,741,897	26,817.0052
	MD	18,527.1646	3,905,472,944	7,021,073,065	3,726,551.68	186,879.0885	4,517,059.75	16,966.594
	SD	9421.82494	1,271,069,390	2,248,014,249	1659,861.04	193,180,811.6	371,472,002	23,502.6252
$F_{14}(x)$	AB	131,154.772	6,816,811.85	2,093,007.76	270,517.596	369,042.9999	333,813.274	38,907.1977
	MD	109,800.311	5,506,707.5	1,692,106.9	248,357.925	137,669.8479	99,365.0474	24,004.7416
	SD	74,612.9136	4,656,160.18	1,375,028.96	113,787.245	640,038.5333	1,180,870.46	40,344.6637
$F_{15}(x)$	AB	8915.09332	519,343,539	512,525,748	522,995.483	58,693.31574	86,213.8881	7888.58976
	MD	5428.6522	428,173,625	475,300,661	514,399.613	35,343.82469	63,456.3421	3430.78723
	SD	12,023.8419	350,472,927	266,935,668	282,908.145	74,805.66654	61,784.9118	8715.94813
$F_{16}(x)$	AB	3371.67432	4820.88686	5247.63056	2937.15125	3160.316909	2823.54207	2647.38414
	MD	3388.42168	4838.74796	5230.02992	2987.83654	3135.128472	2823.76408	2599.69633
	SD	194.740886	282.787939	336.291377	168.503372	330.2217186	407.164035	272.998599

Table 7. *Cont.*

		ABC	CMAES	CSA	DE	MFO	PSO	SOA
$F_{17}(x)$	AB	2395.86777	3482.11031	3378.54815	2177.58318	2441.465302	2294.79893	2174.93599
	MD	2389.4562	3468.22585	3372.19373	2187.84011	2454.470182	2268.3003	2159.79812
	SD	102.889656	285.667211	324.223747	85.3307491	250.2293907	294.2882	194.053384
$F_{18}(x)$	AB	5,046,671.6	37,362,365	23,998,373.2	2,228,087.66	4,549,050.542	1,341,517.26	818,677.628
	MD	4,728,588.97	33,258,799.8	22,232,367.5	2,033,807.45	1,182,780.284	682,252.246	300,231.736
	SD	2,468,179.06	22,078,270.5	13,062,667.1	914,951.248	9,866,226.402	1,857,506.17	1,622,268.27
$F_{19}(x)$	AB	16,681.7816	620,387,919	616,153,747	551,809.083	16,702,987.12	13,973,268.1	4686.90195
	MD	7134.24743	509,584,907	600,010,030	481,336.814	143,655.9903	541,084.259	3334.90276
	SD	24,639.1966	466,495,950	315,926,138	399,361.441	46,487,264.39	44,969,899.6	3974.19763
$F_{20}(x)$	AB	2776.66582	2804.5126	2945.57725	2439.77572	2697.417352	2452.70485	2439.26615
	MD	2764.33673	2834.35184	2955.48441	2445.1877	2627.891978	2488.70557	2449.42599
	SD	94.7983596	199.062724	106.815418	88.0385339	220.4185834	179.132846	147.240337
$F_{21}(x)$	AB	2523.44927	2641.22467	2739.23609	2514.75813	2506.000485	2434.7453	2421.58999
	MD	2524.34506	2643.05906	2741.53087	2518.37232	2499.78219	2433.21201	2426.23287
	SD	15.4046594	36.2479731	41.9080485	10.9762507	40.05305083	23.6970887	21.141776
$F_{22}(x)$	AB	4836.58643	9577.04104	9095.47501	6970.19197	6645.22701	5532.20097	4165.49182
	MD	4242.35722	9932.08496	9075.49459	6750.16911	6848.788542	6523.54829	3767.22155
	SD	2051.83573	1437.66121	763.435864	1310.37051	1508.145086	1932.03672	1801.85242
$F_{23}(x)$	AB	2872.69194	3047.39462	3398.54677	2848.15024	2821.343894	2904.06627	2785.54406
	MD	2872.47313	3045.83859	3425.04431	2848.07645	2814.326953	2917.30159	2781.95288
	SD	10.9306383	35.8015482	80.91846	9.83856624	34.39809682	57.5467798	22.306126
$F_{24}(x)$	AB	3029.76315	3180.2796	3622.26109	3048.38906	2970.576725	3097.14113	2998.32425
	MD	3029.47271	3188.3269	3620.25868	3050.32895	2969.979831	3072.58458	2996.90489
	SD	12.2834187	30.8119037	99.4471973	14.170605	27.40350786	61.1357559	35.0296809
$F_{25}(x)$	AB	2918.77003	3333.22102	6228.94202	2976.74179	3234.960327	3006.6561	2931.4471
	MD	2919.85456	2980.52118	6253.4295	2977.53996	3142.80023	2992.45175	2927.85824
	SD	9.81487821	759.162584	864.95632	16.7265818	310.0654881	102.116534	17.6935953
$F_{26}(x)$	AB	5896.23764	8340.37382	11,171.1937	5643.82962	5776.420284	5702.64017	4464.46857
	MD	5889.23731	8393.26188	11,286.1812	5659.37067	5796.165191	5707.34766	4933.20537
	SD	115.549909	452.740525	720.862035	138.81176	469.5232252	888.194496	945.460049
$F_{27}(x)$	AB	3235.30385	3408.04414	4035.72773	3228.08324	3245.912414	3361.79871	3225.24773
	MD	3235.68049	3407.30733	4021.85789	3228.32551	3242.855656	3349.63975	3224.0298
	SD	6.57578664	33.3663071	190.024017	3.10051948	23.16796867	69.4288619	11.3386465
$F_{28}(x)$	AB	3328.57792	6580.85938	7194.47112	3389.18088	4365.338523	3623.74347	3292.31393
	MD	3326.19	6795.88776	7205.84449	3392.76022	4053.737772	3542.19939	3289.42285
	SD	17.9291186	558.703896	801.137793	29.1593474	920.6971447	251.617478	36.390848
$F_{29}(x)$	AB	4459.90622	5654.19067	6442.47253	4169.15715	4120.329591	4068.18331	3734.31699
	MD	4476.48304	5655.83405	6379.40471	4183.03085	4153.333265	4030.03404	3709.79866
	SD	160.899424	250.599138	525.490575	134.81878	272.9470092	334.451062	167.593361
$F_{30}(x)$	AB	401,261.311	629,251,110	736,794,634	317,868.587	876,056.4301	2,818,096.31	27,288.2215
	MD	343,273.6	484,558,438	795,847,326	289,961.292	213,206.5063	1,323,187.36	23,028.0874
	SD	260,605.287	448,734,951	231,471,227	159,645.774	1,156,943.744	3,490,925.37	12,690.9933

According to Table 7, SOA provides better results than ABC, CMAES, CSA, DE, MFO and PSO for almost all functions. A close inspection of this table reveals that the SOA scheme attained the best performance level, obtaining the best results in 22 functions from the CEC2017 function set. Likewise, the CMAES presents second place, in terms of most of the performance indexes, while DE and CSA techniques reach the third category with performance slightly minor. On the other hand, the MFO and PSO methods produce the worst results. In particular, the results show considerable precision differences, which are directly related to the different search patterns presented by each metaheuristic algorithm. This demonstrates that the search patterns, produced by second-order systems, are able to provide excellent search patterns. The results show good performance of the proposed SOA method in terms of accuracy and high convergence rates.

8. Analysis and Discussion

The extensive experiments, performed in previous sections, demonstrate the remarkable characteristics of the proposed SOA algorithm. The experiments included not only standard benchmark functions but also the complex set of optimization functions from CEC2017. In both sets of functions, they have been solved in 30 dimensions. Therefore, a total of 50 optimization problems were employed to comparatively evaluate the performance of SOA with other popular metaheuristic approaches, such as ABC, CMAES, CSA, DE, MFO and PSO. From the experiments, important information has been obtained by observing the end-results, in terms of the mean and standard deviations found over a certain number of runs or convergence graphs, but also in-depth search behavioral evidence in the form of exploration and exploitation measurements were also used.

The generation of efficient search patterns for the correct exploration of a fitness landscape could be complicated, particularly in the presence of ruggedness and multiple local optima. A search pattern is a set of movements produced by a rule or model that is used to examine promising solutions from the search space. Exploration and exploitation correspond to the most important characteristics of a search strategy. The combination of both mechanisms in a search pattern is crucial for attaining success when solving a particular optimization problem.

In our approach, the temporal response of second-order system is used to generate the trajectory from the position of $\mathbf{x}_i^k = \left\{ x_{i,1}^k, \ldots, x_{i,d}^k \right\}$ to the location of $\mathbf{b} = \{b_1, \ldots, b_d\}$. Three different search patterns have been considered based on the second-order system responses. The proposed search patterns can explore areas of considerable size by using a high rate of velocity and the same time, refining the solution of the best individual $\mathbf{b}$ by the exploitation of its location. This behavior represents the most important property of the proposed search patterns. According to the results provided by the experiments, the search patterns produce more complex trajectories that allow a better examination of the search space.

Similar to other metaheuristic methods, SOA tries to improve its solutions based on its interaction with the objective function or on a 'trial and error' scheme through defined stochastic processes. Different from other popular metaheuristic methods such as DE, ABC, GA or CMAES, our proposed approach uses search patterns represented by trajectories to explore and exploit the search space. Since SOA employs search patterns, it presents more similarities with algorithms such as CSA, MFO and GWO. However, the search patterns used in their search strategy are very different. While CSA, MFO and GWO consider only spiral patterns, our proposed method uses complex trajectories produced by the response of second-order systems.

9. Conclusions

A search pattern is a set of movements produced by a rule or model, in order to examine promising solutions from the search space. In this paper, a set of new search patterns are introduced to explore the search space efficiently. They emulate the response of a second-order system. Under such conditions, it is considered three different responses of a second-order system to produce three distinct search patterns, such as underdamped, critically damped and overdamped. These proposed set of search patterns have been integrated as a complete search strategy, called Second-Order Algorithm (SOA), to obtain the global solution of complex optimization problems.

The form of the search patterns allows for balancing the exploration and exploitation abilities by efficiently traversing the search-space and avoiding suboptimal regions. The efficiency of the proposed SOA has been evaluated through 20 standard benchmark functions and 30 functions of CEC2017 test-suite. The results over multimodal functions show remarkable exploration capabilities, while the result over unimodal test functions denotes adequate exploitation of the search space. On hybrid functions, the results demonstrate the effectivity of the search patterns on more difficult formulations. The search efficacy of the proposed approach is also analyzed in terms of the Wilcoxon test results and convergence curves. In order to compare the performance of the SOA scheme, many other popular optimization techniques such as the Artificial Bee Colony (ABC), the Covariance matrix adaptation evolution strategy (CMAES), the Crow Search Algorithm (CSA), the Differential Evolution (DE), the Moth-flame optimization algorithm (MFO) and the Particle Swarm Optimization (PSO), have also been tested on the same experimental environment. Future research directions include topics such as multi-objective capabilities, incorporating chaotic maps and include acceleration process to solve other real-scale optimization problems.

Author Contributions: Conceptualization, E.C. and M.J.; Data curation, M.P.; Formal analysis, H.E.; Funding acquisition, M.J.; Investigation, H.B. and H.F.E.; Methodology, H.B., A.L.-C. and H.F.E.; Resources, H.E. and M.J.; Supervision, A.L.-C.; Visualization, M.P.; Writing—original draft, E.C. and H.F.E. All authors have read and agreed to the published version of the manuscript.

Funding: This research received no external funding.

Institutional Review Board Statement: Not applicable.

Informed Consent Statement: Not applicable.

Data Availability Statement: Not applicable.

Conflicts of Interest: The authors declare no conflict of interest.

Appendix A

Table 1. List of Benchmark Functions.

	Name	Function	S	Dim	Minimum
$f_1(x)$	Levy	$sin^2(\pi\omega_1) + \sum_{i=1}^{d-1}(\omega_i-1)^2\left[1+10sin^2(\pi\omega_i+1)+(\omega_d-1)^2\left[1+sin^2(2\pi\omega_d)\right]\right]$	$[-10,10]^n$	30	$f(\mathbf{x}^*)=0;$ $\mathbf{x}^*=(1,\ldots,1)$
$f_2(x)$	Mishra 1	$(1+x_n)^{x_n};\quad x_n=n-\sum_{i=1}^{n-1}x_i$	$[0,1]^n$	30	$f(\mathbf{x}^*)=2;$ $\mathbf{x}^*=(1,\ldots,1)$
$f_3(x)$	Mishra 2	$(1+x_n)^{x_n};\quad x_n=n-\sum_{i=1}^{n-1}\frac{(x_i+x_{i+1})}{2}$	$[0,1]^n$	30	$f(\mathbf{x}^*)=2;$ $\mathbf{x}^*=(1,\ldots,1)$
$f_4(x)$	Mishra 11	$\left[\frac{1}{n}\sum_{i=1}^{n}|x_i|-\left(\prod_{i=1}^{n}|x_i|\right)^{\frac{1}{n}}\right]^2$	$[-10,10]^n$	30	$f(\mathbf{x}^*)=0;$ $\mathbf{x}^*=(0,\ldots,0)$
$f_5(x)$	Penalty 1	$\frac{\pi}{30}\left\{\begin{array}{c}10\sin^2(\pi y_1)\\+\sum_{i=1}^{n-1}(y_i-1)^2\left[1+10\sin^2(\pi y_i+1)\right]\\+(y_n-1)^2\end{array}\right\}+\sum_{i=1}^{n}u(x_i,10,100,4);$ $y_i=1+\frac{x_i+1}{4};$ $u(x_i,a,k,m)=\left\{\begin{array}{c}k(x_i-a)^m,\ x_i>a\\0,\ -a\leq x_i\leq a\\k(-x_i-a)^m,\ x_i<-a\end{array}\right.$	$[-50,50]^n$	30	$f(\mathbf{x}^*)=0;$ $\mathbf{x}^*=(-1,\ldots,-1)$
$f_6(x)$	Perm1	$\sum_{k=1}^{n}\left[\sum_{i=1}^{n}(i^k+50)\left\{(x_i/i)^k-1\right\}\right]^2$	$[-n,n]^n$	30	$f(\mathbf{x}^*)=0;$ $\mathbf{x}^*=(1,2,\ldots,n)$
$f_7(x)$	Plateau	$30+\sum_{i=1}^{n}|x_i|$	$[-5.12,5.12]^n$	30	$f(\mathbf{x}^*)=30;$ $\mathbf{x}^*=(0,\ldots,0)$
$f_8(x)$	Step	$\sum_{i=1}^{n}(x_i+0.5)^2$	$[-100,100]^n$	30	$f(\mathbf{x}^*)=0;$ $\mathbf{x}^*=(0,\ldots,0)$
$f_9(x)$	Styblinski tang	$\frac{1}{2}\sum_{i=1}^{n}(x_i^4-16x_i^2+5x_i)$	$[-5,5]^n$	30	$f(\mathbf{x}^*)=-39.1659n;$ $\mathbf{x}^*=(-2.90,\ldots,2.90)$
$f_{10}(x)$	Trid	$\sum_{i=1}^{n}(x_i-1)^2-\sum_{i=1}^{n}x_ix_i-1$	$[-n^2,n^2]^n$	30	$f(\mathbf{x}^*)=$ $-n(n+4)(n-1)/6;$ $\mathbf{x}^*=[i(n+1-i)]$ for $i=1,\ldots,n$
$f_{11}(x)$	Vincent	$-\sum_{i=1}^{n}\sin(10\log x_i)$	$[0.25,10]^n$	30	$f(\mathbf{x}^*)=-n;$ $\mathbf{x}^*=(7.70,\ldots,7.70)$
$f_{12}(x)$	Zakharov	$\sum_{i=1}^{n}x_i^2+(\sum_{i=1}^{n}0.5ix_i)^2+(\sum_{i=1}^{n}0.5ix_i)^4$	$[-5,10]^n$	30	$f(\mathbf{x}^*)=0;$ $\mathbf{x}^*=(0,\ldots,0)$
$f_{13}(x)$	Rothyp	$\sum_{i=1}^{d}\sum_{j=1}^{i}x_j^2$	$[-65.536,65.536]^n$	30	$f(\mathbf{x}^*)=0;$ $\mathbf{x}^*=(0,\ldots,0)$

Table 1. *Cont.*

	Name	Function	S	Dim	Minimum
$f_{14}(x)$	Schwefel 2	$\sum_{i=1}^{n}\left(\sum_{j=1}^{i} x_i\right)^2$	$[-100, 100]^n$	30	$f(\mathbf{x}^*) = 0;$ $\mathbf{x}^* = (0,\dots,0)$
$f_{15}(x)$	Sum2	$\sum_{i=1}^{d}\sum_{j=1}^{i} x_j^2$	$[-10, 10]^n$	30	$f(\mathbf{x}^*) = 0;$ $\mathbf{x}^* = (0,\dots,0)$
$F_{16}(x)$	Sum of different powers	$\sum_{i=1}^{d} \|x_i\|^{i+1}$	$[-1, 1]^n$	30	$f(\mathbf{x}^*) = 0;$ $\mathbf{x}^* = (0,\dots,0)$
$f_{17}(x)$	Rastringin + Schwefel22 + Sphere	$10n + \sum_{i=1}^{n}\left[x_i^2 - 10\cos(2\pi x_i)\right] + \left(\sum_{i=1}^{n}\|x_i\| + \prod_{i=1}^{n}\|x_i\|\right) + \left(\sum_{i=1}^{n} x_i^2\right)$	$[-100, 100]^n$	30	$f(\mathbf{x}^*) = 0;$ $\mathbf{x}^* = (0,\dots,0)$
$f_{18}(x)$	Griewank + Rastringin + Rosenbrock	$\frac{1}{4000}\sum_{i=1}^{n} x_i^2 - \prod_{i=1}^{n}\cos\left(\frac{x_i}{\sqrt{i}}\right) + 1 + 10n + \sum_{i=1}^{n}\left[x_i^2 - 10\cos(2\pi x_i)\right] +$ $\sum_{i=1}^{n-1}\left[100(x_{i+1} - x_i^2)^2 + (x_i - 1)^2\right]$	$[-100, 100]^n$	30	$f(\mathbf{x}^*) = n - 1;$ $\mathbf{x}^* = (0,\dots,0)$
$f_{19}(x)$	Ackley + Penalty2 + Rosenbrock + Schwefel2	$\left(-20exp\left(-0.2\sqrt{\frac{1}{n}\sum_{i=1}^{n} x_i^2}\right) - \exp\left(\frac{1}{n}\sum_{i=1}^{n}\cos(2\pi x_i)\right) + 20 + exp\right) +$ $\left(0.1\left\{\sin(3\pi x_i) + \sum_{i=1}^{n}(x_i - 1)^2\right.\right.$ $\left.\left.[1 + sin^2(3\pi x_i + 1)] + [(x_n - 1)^2[1 + sin^2(2\pi x_n)]]\right\} + \sum_{i=1}^{n} u(x_i, 5.100, 4)\right) +$ $\left(\sum_{i=1}^{n-1}\left[100(x_{i+1} - x_i^2)^2 + (x_i - 1)^2\right]\right) + \left(\sum_{i=1}^{n}\|x_i\| + \prod_{i=1}^{n}\|x_i\|\right)$	$[-100,100]^n$	30	$f(\mathbf{x}^*) = (1.1n) - 1;$ $\mathbf{x}^* = (0,\dots,0)$
$f_{20}(x)$	Ackley + Griewnk + Rastringin + Rosenbrock + Schwefel22	$-20e^{-0.2\sqrt{\frac{1}{n}\sum_{i=1}^{n} x_i^2}} - e^{\frac{1}{n}\sum_{i=1}^{n}\cos(2\pi x_i)} + 20 + e + \frac{1}{4000}\sum_{i=1}^{n} x_i^2 -$ $\prod_{i=1}^{n}\cos\left(\frac{x_i}{\sqrt{i}}\right) + 1 + 10n + \sum_{i=1}^{n}\left[x_i^2 - 10\cos(2\pi x_i)\right] +$ $\sum_{i=1}^{n-1}\left[100(x_{i+1} - x_i^2)^2 + (x_i - 1)^2\right] + \sum_{i=1}^{n}\|x_i\| + \prod_{i=1}^{n}\|x_i\|$	$[-100, 100]^n$	30	$f(\mathbf{x}^*) = n - 1;$ $\mathbf{x}^* = (0,\dots,0)$

References

1. Cuevas, E.; Gálvez, J.; Avila, K.; Toski, M.; Rafe, V. A new metaheuristic approach based on agent systems principles. *J. Comput. Sci.* **2020**, *47*, 101244. [CrossRef]
2. Beyer, H.-G.; Schwefel, H.-P. Evolution strategies—A comprehensive introduction. *Nat. Comput.* **2002**, *1*, 3–52. [CrossRef]
3. Bäck, T.; Hoffmeister, F.; Schwefel, H.-P. A survey of evolution strategies. In Proceedings of the Fourth International Conference on Genetic Algorithms, San Diego, CA, USA, 13–16 July 1991; p. 8.
4. Hansen, N. The CMA Evolution Strategy: A Tutorial. *arXiv* **2016**, arXiv:1604.00772102, 75–102.
5. Tang, K.S.; Man, K.F.; Kwong, S.; He, Q. Genetic algorithms and their applications. *IEEE Signal Process. Mag.* **1996**, *13*, 22–37. [CrossRef]
6. Storn, R.; Price, K. Differential Evolution—A Simple and Efficient Heuristic for global Optimization over Continuous Spaces. *J. Glob. Optim.* **1997**, *11*, 341–359. [CrossRef]
7. Zhang, J.; Sanderson, A.C. JADE: Self-adaptive differential evolution with fast and reliable convergence performance. In Proceedings of the 2007 IEEE Congress on Evolutionary Computation, CEC 2007, Singapore, 25–28 September 2007.
8. Askarzadeh, A. A novel metaheuristic method for solving constrained engineering optimization problems: Crow search algorithm. *Comput. Struct.* **2016**, *169*, 1–12. [CrossRef]
9. Karaboga, D.; Gorkemli, B.; Ozturk, C.; Karaboga, N. A comprehensive survey: Artificial bee colony (ABC) algorithm and applications. *Artif. Intell. Rev.* **2014**, *42*, 21–57. [CrossRef]
10. Kennedy, J.; Eberhart, R. Particle swarm optimization. In Proceedings of the ICCN'95 International Conference on Neural Networks, Perth, WA, Australia, 27 November–1 December 1995; pp. 1942–1948.
11. Poli, R.; Kennedy, J.; Blackwell, T. Particle swarm optimization. *Swarm Intell.* **2007**, *1*, 33–57. [CrossRef]
12. Marini, F.; Walczak, B. Particle swarm optimization (PSO): A tutorial. *Chemom. Intell. Lab. Syst.* **2015**, *149*, 153–165. [CrossRef]
13. Yang, X.-S. Firefly Algorithms for Multimodal Optimization. In Proceedings of the 5th International Conference on Stochastic Algorithms: Foundations and Applications, Sapporo, Japan, 26–28 October 2009; Springer: Berlin/Heidelberg, Germany, 2009; pp. 169–178. [CrossRef]
14. Yang, X.-S. Firefly Algorithm, Lévy Flights and Global Optimization. In *Research and Development in Intelligent Systems XXVI*; Springer: London, UK, 2010; pp. 209–218. [CrossRef]
15. Yang, X.S.; Deb, S. Cuckoo search via Lévy flights. In Proceedings of the 2009 World Congress on Nature & Biologically Inspired Computing (NaBIC), Coimbatore, India, 9–11 December 2009; pp. 210–214.
16. Yang, X.-S. A new metaheuristic Bat-inspired Algorithm. *Stud. Comput. Intell.* **2010**, *284*, 65–74.
17. Mirjalili, S.; Mirjalili, S.M.; Lewis, A. Grey Wolf Optimizer. *Adv. Eng. Softw.* **2014**, *69*, 46–61. [CrossRef]
18. Mirjalili, S. Moth-flame optimization algorithm: A novel nature-inspired heuristic paradigm. *Knowl. Based Syst.* **2015**, *89*, 228–249. [CrossRef]
19. Cuevas, E.; Echavarría, A.; Ramírez-Ortegón, M.A. An optimization algorithm inspired by the States of Matter that improves the balance between exploration and exploitation. *Appl. Intell.* **2013**, *40*, 256–272. [CrossRef]

20. Valdivia-Gonzalez, A.; Zaldívar, D.; Fausto, F.; Camarena, O.; Cuevas, E.; Perez-Cisneros, M. A States of Matter Search-Based Approach for Solving the Problem of Intelligent Power Allocation in Plug-in Hybrid Electric Vehicles. *Energies* **2017**, *10*, 92. [CrossRef]
21. Kirkpatrick, S.; Gelatt, C.D.; Vecchi, M.P. Optimization by Simulated Annealing. *Science* **1983**, *220*, 671–680. [CrossRef]
22. Rutenbar, R.A. Simulated Annealing Algorithms: An Overview. *IEEE Circuits Devices Mag.* **1989**, *5*, 19–26. [CrossRef]
23. Siddique, N.; Adeli, H. Simulated Annealing, Its Variants and Engineering Applications. *Int. J. Artif. Intell. Tools* **2016**, *25*, 1630001. [CrossRef]
24. Rashedi, E.; Nezamabadi-Pour, H.; Saryazdi, S. GSA: A Gravitational Search Algorithm. *Inf. Sci.* **2009**, *179*, 2232–2248. [CrossRef]
25. Eskandar, H.; Sadollah, A.; Bahreininejad, A.; Hamdi, M. Water cycle algorithm—A novel metaheuristic optimization method for solving constrained engineering optimization problems. *Comput. Struct.* **2012**, *110–111*, 151–166. [CrossRef]
26. Erol, O.K.; Eksin, I. A new optimization method: Big Bang–Big Crunch. *Adv. Eng. Softw.* **2006**, *37*, 106–111. [CrossRef]
27. Birbil, Ş.I.; Fang, S.C. An electromagnetism-like mechanism for global optimization. *J. Glob. Optim.* **2003**, *25*, 263–282. [CrossRef]
28. Sörensen, K.; Glover, F. Metaheuristics. In *Encyclopedia of Operations Research and Management Science*; Gass, S.I., Fu, M., Eds.; Springer: New York, NY, USA, 2013; pp. 960–970.
29. Zill, D.G. *A First Course in Differential Equations with Modeling Applications*; Cengage Learning: Boston, MA, USA, 2012; ISBN 978-1-285-40110-2.
30. Haidekker, M.A. *Linear Feedback Controls*; Elsevier: Amsterdam, The Netherlands, 2013.
31. Morales-Castañeda, B.; Zaldívar, D.; Cuevas, E.; Fausto, F.; Rodríguez, A. A better balance in metaheuristic algorithms: Does it exist? *Swarm Evol. Comput.* **2020**, *54*, 100671. [CrossRef]
32. Boussaïda, I.; Lepagnot, J.; Siarry, P. A survey on optimization metaheuristics. *Inf. Sci.* **2013**, *237*, 82–117. [CrossRef]
33. Han, M.; Liu, C.; Xing, J. An evolutionary membrane algorithm for global numerical optimization problems. *Inf. Sci.* **2014**, *276*, 219–241. [CrossRef]
34. Meng, Z.; Pan, J.-S. Monkey King Evolution: A new memetic evolutionary algorithm and its application in vehicle fuel consumption optimization. *Knowl. Based Syst.* **2016**, *97*, 144–157. [CrossRef]
35. Wilcoxon, F. Individual comparisons by ranking methods. *Biometrics* **1945**, 80–83. [CrossRef]
36. Wu, G.H.; Mallipeddi, R.; Suganthan, P.N. Problem Definitions and Evaluation Criteria for the CEC 2017 Competition on Constrained Real-Parameter Optimization. Available online: https://www.researchgate.net/profile/Guohua-Wu-5/publication/317228117_Problem_Definitions_and_Evaluation_Criteria_for_the_CEC_2017_Competition_and_Special_Session_on_Constrained_Single_Objective_Real-Parameter_Optimization/links/5982cdbaa6fdcc8b56f59104/Problem-Definitions-and-Evaluation-Criteria-for-the-CEC-2017-Competition-and-Special-Session-on-Constrained-Single-Objective-Real-Parameter-Optimization.pdf (accessed on 16 September 2019).

Article

A Feature-Independent Hyper-Heuristic Approach for Solving the Knapsack Problem

Xavier Sánchez-Díaz, José Carlos Ortiz-Bayliss *, Ivan Amaya, Jorge M. Cruz-Duarte, Santiago Enrique Conant-Pablos and Hugo Terashima-Marín

Tecnologico de Monterrey, School of Engineering and Sciences, Monterrey 64849, Mexico; sax@tec.mx (X.S.-D.); iamaya2@tec.mx (I.A.); jorge.cruz@tec.mx (J.M.C.-D.); sconant@tec.mx (S.E.C.-P.); terashima@tec.mx (H.T.-M.)
* Correspondence: jcobayliss@tec.mx

Abstract: Recent years have witnessed a growing interest in automatic learning mechanisms and applications. The concept of hyper-heuristics, algorithms that either select among existing algorithms or generate new ones, holds high relevance in this matter. Current research suggests that, under certain circumstances, hyper-heuristics outperform single heuristics when evaluated in isolation. When hyper-heuristics are selected among existing algorithms, they map problem states into suitable solvers. Unfortunately, identifying the features that accurately describe the problem state—and thus allow for a proper mapping—requires plenty of domain-specific knowledge, which is not always available. This work proposes a simple yet effective hyper-heuristic model that does not rely on problem features to produce such a mapping. The model defines a fixed sequence of heuristics that improves the solving process of knapsack problems. This research comprises an analysis of feature-independent hyper-heuristic performance under different learning conditions and different problem sets.

Keywords: hyper-heuristics; knapsack problem; optimization

Citation: Sánchez-Díaz, X. Ortiz-Bayliss, J.C.; Amaya, I.; Cruz-Duarte, J.M.; Conant-Pablos, S.E.; Terashima-Marín, H. A Feature-Independent Hyper-Heuristic Approach for Solving the Knapsack Problem. *Appl. Sci.* **2021**, *11*, 10209. https://doi.org/10.3390/app112110209

Academic Editor:Peng-Yeng Yin

Received: 24 September 2021
Accepted: 26 October 2021
Published: 31 October 2021

Publisher's Note: MDPI stays neutral with regard to jurisdictional claims in published maps and institutional affiliations.

1. Introduction

The intersection between combinatorial optimization and Hyper-Heuristics (HHs) is a relevant and active area in literature, as Sánchez et al. detailed with their thorough systematic review [1]. The former considers optimization problems where a permutation of feasible values gives candidate solutions. The hardness of these problems could be easy or hard to solve, depending on different parameters. Among them resides a category known as NP-hard problems, for which analytical solvers cannot be scaled due to an excessive computational burden. Therefore, approximate solvers are commonly sought for NP-hard problems, including those based on heuristics. It is here where HHs shine bright. This approach has been deemed as high-level heuristics useful to tackle hard-to-solve problems [2], particularly NP-hard ones [3]. A classification of HHs considers two main groups: selection HHs (those that select heuristics from an available set) and generation HHs (those that create new heuristics using components of existing ones) [4]. Although both groups have proved of great interest to the scientific community, in this work we focus on the former.

Selection HHs deal with problems indirectly. They browse a set of available heuristics, which are selectively applied to solve the problem at hand [5]. A selection HH analyzes a set of available heuristics and chooses the most suitable one according to a given performance metric. Most of the current selection HH models include two key phases: heuristic selection and move acceptance [6]. The former represents the strategy for deciding which heuristic should be selected. Conversely, the latter determines whether the new solution is accepted or discarded. The approach proposed in this work simplifies the overall model by only focusing on heuristic selection. Thus, changes resulting from applying a particular heuristic are always accepted.

Evolutionary computation is a recurrent approach among the many learning methods employed in the literature to produce HHs. Some examples include, but are not limited to, Genetic Programming (GP) [5,7–9], Grammatical Evolution (GE) [10], Genetic Algorithms (GA) [11,12], and Artificial Immune Systems (AIS) [13,14]. The literature contains other related proposals, such as those that evolve HHs by analyzing the set of problem instances [15]. These findings support the idea of using an evolutionary strategy as a learning mechanism for HHs, one which improves their performance via crossover or mutation. This work focuses on the latter.

HHs have been used to tackle packing problems in the past. Hart and Sim describe a variant of AIS used as a hybrid HH for the Bin Packing Problem [13,14,16]. Falkenauer [11] proposed a GA-based HH model, and Burke et al. [17], Hyde [8], and Drake et al. [9,18,19] have studied GP rules for the Bin Packing and Multidimensional Knapsack problems. More recent studies of HHs have been conducted on the binary knapsack domain using Ant Colony Optimization [20] and Quartile-related Information [21]. Further information about the state-of-the-art of HHs and their applications are provided in [4,22].

As mentioned above, several combinatorial optimization problems have been tackled with HHs [1]. This paper focuses on the Knapsack Problem (KP), which has been studied in great depth due to its simple structure and broad applicability. Example applications include cargo loading, cutting stock, allocation, and cryptography [23]. When a constructive approach is used to solve this problem, the solution is built one step at a time by deciding if one particular item must be packed or ignored. For simplicity, our setting states that once the process chooses an item, there is no way to change such a decision. Using a constructive approach leads to different subproblems throughout the solving process, depending on the heuristic used. This happens because packing an item produces an instance with a reduced knapsack capacity (the previous knapsack capacity minus the weight of the packed item) and a reduced list of items (the item recently packed or ignored is no longer part of the items to pack). This behavior raises the question of which heuristic, among the different options, should be used to maximize the overall profit resulting from the items contained in the knapsack. The literature usually refers to the problem of selecting the best algorithm for a particular situation as the Algorithm Selection Problem [24].

Despite the extensive use of selection HHs [4,25], only a few works explore the insights of their behavior. A few examples include a run-time analysis [26,27], the use and control of crossover operators [19], and heuristic interaction when applied to constraint satisfaction problems [28]. Furthermore, traditional selection HH models, that represent the relation between problem states and heuristics through ⟨condition, action⟩ rules, exhibit a significant limitation: they require the definition of a set of features to characterize the problem state [29]. Finding such a set of features implies an additional layer of complexity to the model. To the best of our knowledge, there is only one work on feature-independent HHs for the knapsack problem, which obtained only a few preliminary results [30]. Therefore, our work aims at filling this knowledge gap through two particular contributions:

- It proposes an evolutionary-powered hyper-heuristic framework capable of combining the strengths of individual heuristics when solving sets of KP instances.
- Besides the model itself, it provides the analysis and understanding of the performance of the hyper-heuristics designed under this model on instance sets formed by challenging instances, both balanced and unbalanced (in terms of heuristic performance).

The remainder of this document is organized as follows. Section 2 defines the fundamental ideas associated with our work. Section 3 describes the HH model and the rationale behind it. Section 4 details the experiments conducted and analyzes the results. Finally, Section 5 presents the conclusions and provides an overview of future directions for this investigation.

2. Background

This section introduces several basic concepts utilized in this work. It starts by describing the fundamental Knapsack Problem (KP). Then, it presents heuristics in general. Later on, it explains the basic foundations of Evolutionary Algorithms (EAs).

2.1. Knapsack Problem

In layman's terms, a knapsack problem seeks to store a set of items into a knapsack with limited capacity. Therefore, one must choose a subset of the items based on some criteria. For example, the profit or weight to select the items that must go into the knapsack. In formal terms, let $\vec{x} \in \mathbb{Z}_2^n$ be a binary-valued vector of n items, where each element $x_i \in \vec{x}$ represents an item and its selection according to its position and value, respectively, for example, if there is a stock of three items and we choose the second one, we have $\vec{x} = (0, 1, 0)^\mathsf{T}$. Likewise, let $\vec{p} \in \mathbb{R}_+^n$ and $\vec{w} \in \mathbb{R}_+^n$ be the profit and weight vectors directly related to the items, i.e., the i-th item has profit $p_i \in \vec{p}$ and weight $w_i \in \vec{w}$. Thus, this problem (widely known as the 0/1 Knapsack Problem) can be defined as

$$
\begin{aligned}
\vec{x}_* &= \underset{\vec{x}}{\mathrm{argmax}}\{\vec{p} \cdot \vec{x}\}, \\
\text{s.t.} \quad &\vec{w} \cdot \vec{x} \leq c, \\
\text{with} \quad &\vec{x} \in \mathbb{Z}_2^n \quad \text{and} \quad \vec{p}, \vec{w} \in \mathbb{R}_+^n
\end{aligned}
\tag{1}
$$

where $c \in \mathbb{R}_+$ is the total capacity of the knapsack.

Although many solving strategies can be found in the literature, the initial solvers for the knapsack problem were initially divided into two categories: exact methods and approximate ones. Exact methods provide optimal solutions but have limited applications due to their consumption of computational resources [31,32]. Conversely, approximated methods make some assumptions to simplify the solving process to give usually suboptimal solutions [33].

However, the advance in computing power and creativity has lead to extensive diversity of techniques that can be found nowadays, most of them hybrids. To provide a glance at such diversity, we mention some representative works. For example, Genetic Algorithms (GA) and Particle Swarm Optimization (PSO) have been used for solving the multidimensional KP [34,35]. Gómez et al. proposed a Binary Particle Swarm strategy with a genetic operator, also for the multidimensional KP [36]. Another example of recent solving methods includes the one by Razavi and Sajedi, where the authors proposed using Gravitational Search (GSA) for solving the 0/1 KP [37]. Inspired by Quantum Computing, Patvardhan et al. proposed a novel method to solve the KP by using a Quantum-Inspired Evolutionary Algorithm [38]. Modifications to apparently simple ideas can also be found nowadays. For example, Lu et al. solve the 0/1 KP by using Greedy Degree and Expectation Efficiency (GDEE) [39]. Some probabilistic models include Cohort Intelligence (CI) [40] and hybrid heuristics. For example, by using Mean Field Theory [41], Banda et al. generated a probabilistic model capable of replacing a complex distribution of items with an easier one.

The literature on the KP includes a wide variety of solution methods. For a thorough treatment, we refer the reader to the textbooks found in [42,43]. Broadly speaking, the solution methods are of three types: exact methods, heuristic methods, and approximation algorithms (which are special heuristic methods that yield solutions of a guaranteed quality). The best exact methods are now capable of solving instances with thousands of items to proven optimality in reasonable computing times [32,44–46]. However, it is possible to devise instances that are very challenging for exact methods [45]. For this reason, heuristics remain of interest.

2.2. Heuristics

A heuristic is a procedure that creates or modifies a candidate solution for a given problem instance. There are many classifications of heuristics in the literature. Most of them relate to combinatorial optimization domains [47]. In this work, we use the term 'heuristic' to refer to the low-level operations to apply to a problem instance [4]. An illustrative example is a specific way to organize a knapsack based on item size. In mathematical terms, a heuristic $h : \mathcal{X} \mapsto \mathcal{X}$ applied to an instance problem $\mathcal{X}$, may use the current candidate solution $\vec{x} \in \mathcal{X}$, and delivers a new candidate $\vec{x}'$ as shown,

$$\vec{x}' = h\{\vec{x}\}. \tag{2}$$

Consider that if $\vec{x} = \varnothing$, the heuristic creates a candidate solution; otherwise, it modifies the current one.

A HH is described as a high-level strategy that controls heuristics in the process of solving an optimization problem [47]. Therefore, HHs move in the heuristic space to find a heuristic configuration that solves a given problem. With that in mind, a HH can be defined as follows [4]. Let $h \in \mathfrak{H}^{\omega}$ be a heuristic sequence from the heuristic space $\mathfrak{H}$, and let $F(h|\mathcal{X}) : \mathfrak{H}^{\omega} \times \mathcal{X} \mapsto \mathbb{R}$ be its performance measure function. Here, $\mathcal{X}$ is the problem domain in an optimization problem with an objective function $f(\vec{x}) : \mathcal{X} \mapsto \mathbb{R}$. For the particular case of the knapsack problem, $\mathcal{X} = \mathbb{Z}_2^n$. Then, a solution $\vec{x}_* \in \mathcal{X}$ and its corresponding fitness value $f(\vec{x}_*)$ are found when a h is applied on $\mathcal{X}$, so its performance $F(h|\mathcal{X})$ can also be determined. Therefore, let a HH be a technique that solves

$$(h_*; \vec{x}_*) = \underset{h \in \mathfrak{H}^{\omega}, \vec{x} \in \mathcal{X}}{\mathrm{argmax}} \{F(h|\mathcal{X})\}. \tag{3}$$

In other words, a HH searches for the optimal heuristic configuration h_* that produces the optimal solution $\vec{x}_*$ with the maximal performance $F(h_*|\mathcal{X})$.

It is essential to mention that there also exists something we might categorize as mid-level heuristics: Metaheuristics (MHs). These methods are trendy because of their proven performance on different scenarios and applications [48]. MHs are defined as master strategies that control simple heuristics [49]. Therefore, by finding a middle point between definitions for low and high-level heuristics, it is possible to say that an MH corresponds to a heuristic sequence h applied recurrently until reaching the desired performance. This idea has been extended by Cruz-Duarte et al. [50,51]. Thus, for a given optimization problem, we can define the process of approximating the optimal solution $\vec{x}_* \in \mathcal{X}$ via an MH such as

$$\vec{x}_* \triangleq \overset{h_f}{\underset{h_k \in h}{\mathbb{M}}} h_k\{\mathcal{X}\} \tag{4}$$

where h_f is also an operator to evaluate stopping criteria, and $\mathbb{M}$ is the iteral operator based on that in [52]. This operator indicates a recurrent application of heuristics from a sequence h until h_f marks the halt. For example, for a MH defined with two operations ($\omega = 2$), say crossover (h_1) and mutation (h_2), thence $h = \{h_1, h_2\}$. Consider h_f as a common fitness tolerance criterion such as $f(\vec{x}) \leq \varepsilon$. Therefore, the metaheuristic applies first h_1, followed by h_2, and then it checks the condition given by h_f. If such a condition is not met, the process is repeated until it complies.

2.3. Evolutionary Algorithms (EAs)

EAs are a particular subgroup of the population-based metaheuristics inspired by evolution and biological processes observed in nature [53]. The most notable examples of these methods are Differential Evolution (DE) [54] and Genetic Algorithms (GAs) [55]. In a general sense, the individuals of an EA interact with each other to explore the problem domain and exploit the candidate solutions, i.e., they evolve. Such evolution is possible due to operators such as selection, crossover, mutation, and reproduction. It is easy to notice that these operators are just (low-level) heuristics. In this work, we chiefly employ mutation-based operations, which are detailed in the next section.

3. Proposed Hyper-Heuristic Model

The HH model developed in this work can be classified as an offline selection HH [56]. Internally, the HH model relies on a variant of the $(1 + 1)$ Evolutionary Algorithm (EA) to find the sequence of heuristics to apply. The original EA flipped each bit with probability $1/\varpi$ (where ϖ is the chromosome length). Conversely, our approach chooses one among various available mutation operators based on a uniform random probability distribution. This EA implementation considers some of the features used by Lehre and Özcan [27]. However, the set of available operators is different as we work with constructive heuristics while they have used perturbative ones.

We choose four simple packing heuristics due to their popularity: $\mathfrak{H} = \{$Def, MaxP, MaxPW, MinW$\}$. Before moving forward, we describe them below.

Default (Def) packs the items in the same order that they are contained in the instance, as long as they fit in the knapsack.

Maximum profit (MaxP) sorts the items in descending order based on their profit and packs them in that order as long as they fit in the knapsack.

Maximum profit per weight (MaxPW) calculates profit-over-weight ratio for each item and sorts them in descending order. Then, MaxPW follows this ordering until the knapsack is full or no more items are left to be packed.

Minimum weight (MinW) favors lighter items, so it sorts items in ascending order based on their weight and packs them by following such order until they no longer fit.

Should there be a tie, all heuristics will choose the first conflicting item. Among these heuristics, Def is the fastest one to execute as it involves no additional operations. On the contrary, MaxP, MaxPW, and MinW take longer to compute but usually yield better results than using no order at all (Def). We are aware that more complex heuristics are available in the literature. Still, at this point in the investigation, we consider it suitable to test the proposed HH approach on this set of simple heuristics.

In our model, each HH represents a sequence of heuristics to apply to the problem in one specific order. For simplicity, we refer to its length as the cardinality of the HH. Let us consider the HH with cardinality of five ($h \in \mathfrak{H}^5$) given by $h = \{$Def, MaxP, MaxP, Def, MinW$\}$ (Figure 1). Please note that one of the available heuristics is not contained within the HH, MaxPW. In this HH, Def occupies positions 0 and 3, MaxP occupies positions 1 and 2, and MinW occupies position 4. Let us assume that h will solve a 12-item knapsack instance. In this case, there are 12 decisions to be made for solving this instance (d_1 to d_{12}). While some of the decisions will add an item to the knapsack, others will discard it. As the HH represents a sequence of five heuristics, the first five decisions will be made in the same order in which the heuristics are presented in h: Def, MaxP, MaxP, Def, and MinW. After that, there are no more heuristics to choose to make further decisions. Then, the HH is used again, but now backwards: MinW, Def, MaxP, MaxP, and Def. The process repeats by inverting the sequence every time we reach the end of the HH. Although this scheme seems disruptive, we consider it feasible to favor the changes in heuristics throughout the solving process.

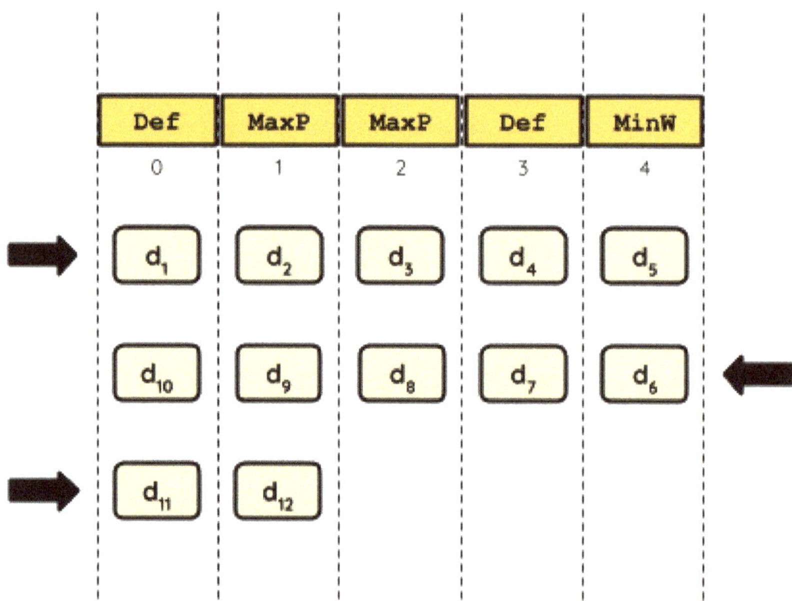

Figure 1. An example of a hyper-heuristic with a cardinality of five, and the way it solves an instance with 12 items. Black arrows indicate the direction in which heuristics are selected to make a decision.

3.1. Hyper-Heuristic Training

The learning mechanism within the HH model works as follows. The HH, $h \in \mathfrak{H}^\omega$, is randomly initialized (with a sequence of ω randomly selected heuristics), and it is used to solve the set of training instances. We register the profit of the solution obtained by h for each instance. The performance of the HH, $F(h, \mathfrak{X})$, is then calculated as the average of all the profits obtained for the training set. In our EA implementation, a chromosome represents a HH that codes a sequence of heuristics to apply, i.e., h. At each iteration, the process randomly chooses one mutation operator o_m from the available operators $\mathfrak{O}$. To do so, a uniform probability distribution is employed. Thus, EA applies it on a copy of the current HH, which produces a candidate HH, according to (2). The model evaluates the candidate HH on the set of training instances. If its performance is greater or equal (to favor diversity) than the current HH, this candidate takes its place. The process is repeated until it reaches the maximum number of allowed iterations $I_{\max}$. Pseudocode 1 details this training procedure.

The main goal of the learning mechanism is to produce a HH (an ordered sequence of heuristics) that maximizes the profit of an instance by packing appropriate items into the knapsack. In other words, it solves (1). Thus, the EA does not operate on the solution space $\mathfrak{X}$, but on the heuristic space $\mathfrak{H}$.

Pseudocode 1 Hyper-heuristic trainer

---

input:
    Set of heuristics $\mathfrak{H}$,
    Set of mutation operators $\mathfrak{O}$,
    Set of instances to train the hyper-heuristic $\mathfrak{X}$,
    Initial cardinality ω,
    Performance function $F(\boldsymbol{h}, \mathfrak{X})$, and
    Maximum number of iterations $I_{\max}$
output: Best selection hyper-heuristic $\boldsymbol{h}_*$

```
 1: h ← INITIALIZERANDOMLY(𝔥, ω)
 2: F_h ← F(h, 𝔛)                           ▷ Evaluate the current HH
 3: for i = 0 to I_max do
 4:     h' ← GENERATECANDIDATE(h)
 5:     F_h' ← F(h', 𝔛)                     ▷ Evaluate the candidate HH
 6:     if F_h' ≥ F_h then
 7:         h ← h' and F_h ← F_h'
 8:     end if
 9: end for
10: return h_* ← h

11: procedure GENERATECANDIDATE(h)
12:     h' ← CLONE(h)
13:     o_m ← CHOOSERANDOMLY(𝔒)
14:     h' ← o_m{h'}                        ▷ Mutate h' by applying o_m
15:     return h'
16: end procedure
```

3.2. Mutation Operators

The evolutionary process dynamically chooses among eight available mutation operators to alter the candidate hyper-heuristic $\boldsymbol{h}'$. Operators contained within set $\mathfrak{O}$ are described below. To clarify the behavior of such operators, we also provide a brief example of their behavior. Bear in mind that, in all the examples, we always consider the HH depicted in Figure 1 as the HH to mutate, i.e., $\boldsymbol{h}' = \{\texttt{Def}, \texttt{MaxP}, \texttt{MaxP}, \texttt{Def}, \texttt{MinW}\}$.

Add inserts a randomly chosen heuristic at a random position $i \sim \mathcal{U}\{0, \omega - 1\}$ into $\boldsymbol{h}'$. In doing so, cardinality ($\omega = \#\boldsymbol{h}'$) increases by one and existing heuristics at this position onward are shifted to the right. For the example, let us assume that $i = 3$ and that the random selection chooses $\texttt{MaxPW}$. Thus, the resulting HH has a cardinality of six and is given by $\boldsymbol{h}' = \{\texttt{Def}, \texttt{MaxP}, \texttt{MaxP}, \texttt{MaxPW}, \texttt{Def}, \texttt{MinW}\}$.

Single-point Flip selects a position $i \sim \mathcal{U}\{0, \omega - 1\}$ at random from $\boldsymbol{h}'$ and changes its heuristic $h_i \in \boldsymbol{h}$ to a different one (selected at random). Thus, cardinality is preserved. Let us suppose that $i = 0$. As $\texttt{Def}$ is the current heuristic at this position, it cannot be chosen. Therefore, let us assume that the new heuristic is $\texttt{MinW}$. Then, the resulting HH is $\boldsymbol{h}' = \{\texttt{MinW}, \texttt{MaxP}, \texttt{MaxP}, \texttt{Def}, \texttt{MinW}\}$.

Two-point Flip is a more disruptive version of the previous operator. This time, heuristics at two different positions ($i, j \sim \mathcal{U}\{0, \omega - 1\}$) are renewed. Let us suppose that $i = 2$ and $j = 4$. In the first case, the available heuristics are $\texttt{Def}$, $\texttt{MaxPW}$, and $\texttt{MinW}$. In the second one, they are $\texttt{Def}$, $\texttt{MaxP}$, and $\texttt{MaxPW}$. Imagine that $\texttt{MaxPW}$ and $\texttt{Def}$ are selected, respectively. Then, the resulting HH still preserves cardinality and is given by $\boldsymbol{h}' = \{\texttt{Def}, \texttt{MaxP}, \texttt{MaxPW}, \texttt{Def}, \texttt{Def}\}$.

Neighbor-based Add is similar to $\texttt{Add}$ as it also inserts a heuristic at a random position $k \sim \mathcal{U}\{0, \omega - 1\}$ within $\boldsymbol{h}'$. However, the heuristic to insert is randomly selected among neighboring heuristics (positions $j = i - 1$ and $k = i + 1$), as long as they are valid positions. Should i correspond to an edge, the only neighbor is copied. Let us suppose that $i = 1$. In this case, the heuristic to insert would be either Def ($j = 0$) or MaxP ($k = 2$), with the same probability (50%). Imagine that MaxP is selected. Then, and as with the Add operator, cardinality grows to six and the resulting HH is $h' = \{\text{Def}, \text{MaxP}, \text{MaxP}, \text{MaxP}, \text{Def}, \text{MinW}\}$.

Neighbor-based Single-point Flip is a variant of Single-Point Flip where the heuristic at $i \sim \mathcal{U}\{0, \omega - 1\}$ changes into one of its neighbors (positions $j = i - 1$ and $k = i + 1$, if they exist). For example, suppose that $i = 2$. In this case, candidate replacements would be either MaxP ($j = 1$) or Def ($k = 3$), each one with a 50% probability. Let us suppose that Def is selected. Then, the resulting HH preserves cardinality and becomes $h' = \{\text{Def}, \text{MaxP}, \text{Def}, \text{Def}, \text{MinW}\}$.

Neighbor-based Two-point Flip likewise, this is a variant of Two-Point Flip in which heuristics are chosen at random from neighboring ones. Let us suppose that $i = 2$ and $j = 4$. In the first case, available heuristics are MaxP and Def; in the second one, it is only Def since the position corresponds to an edge of the HH. Let us suppose that, in the first case, Def is selected. Therefore, the resulting HH is $h' = \{\text{Def}, \text{MaxP}, \text{Def}, \text{Def}, \text{Def}\}$.

Swap interchanges heuristics at two randomly selected locations, $i, j \sim \mathcal{U}\{0, \omega - 1\}$, and so preserves cardinality. For example, assume that $i = 1$ and $j = 3$. Thus, the resulting HH becomes $h' = \{\text{Def}, \text{Def}, \text{MaxP}, \text{MaxP}, \text{MinW}\}$.

Remove randomly selects one position $i \sim \mathcal{U}\{0, \omega - 1\}$ within the HH and removes the heuristic at that position. Therefore, cardinality is reduced by one. After removing the heuristic, upcoming ones are shifted to the left. For example, imagine that $i = 0$. Then, the resulting HH has a cardinality of four and is given by: $h' = \{\text{MaxP}, \text{MaxP}, \text{Def}, \text{MinW}\}$.

Note that neighbor-based operators are allowed to replace the corresponding heuristic with the same one since it is determined by the neighbors. Certainly, other operators can be used but, for the sake of simplicity, we limit ourselves to these eight to formulate $\mathfrak{O}$. Finally, consider that there is only one operator that reduces cardinality (Remove), while two of them increase it (Add and Neighbor-based Add), and the remaining five preserve it (Single-point Flip, Two-point Flip, Single-point Neighbor-based Flip, Two-point Neighbor-based Flip, and Swap). Therefore, it is rather expected that trained HHs exhibit a higher cardinality than untrained ones. Because of this, a HH may end up choosing a different amount of items with each heuristic. We believe such flexibility in the learning process may favor more complex interactions between the available heuristics.

3.3. The Knapsack Problem Instances

In this investigation, we consider four datasets: α_1, α_2, β_1, and β_2. Dataset α_1 is used for training purposes and comprises 400 knapsack problem instances. We synthetically generated such instances for this work by using the algorithm proposed by Plata et al. [57]. This dataset contains a balanced mixture of problem instances, where each heuristic has an equal probability of performing best. Therefore, we are sure that no single heuristic outperforms others on all instances. We replicated the process to produce 400 additional problem instances for testing purposes (dataset α_2). These two datasets consider instances of 50 items and a knapsack capacity of 20 units. Dataset β_1 contains 600 hard instances proposed by Pisinger [45] and featuring 20 items but at different capacities. Finally, dataset β_2 consists of 600 additional hard instances (also proposed by Pisinger [45]). In this case, instances have 50 items and, again, exhibit different capacities. A summary of the datasets is provided in Table 1. Bear in mind that for the confirmatory experiments, we analyze the effect of changing the training dataset, but we shall discuss it in more detail further on.

Table 1. Datasets used in this work.

ID	Type	Features		
		Instances	Items	Capacity
α_1	Balanced	400	50	50
α_2	Balanced	400	50	50
β_1	Hard	600	20	Variable
β_2	Hard	600	50	Variable

3.4. Performance Metrics

We evaluate performance by considering both absolute and relative performances. All approaches are evaluated based on the profits obtained by their solutions (the sum of the profits of the items packed within the knapsack). The overall performance of a method on a particular set is calculated as the sum of the profits of the solutions produced for all the instances in such a set. However, it is also useful to estimate the relative performance, i.e., w.r.t. the other methods. For this purpose, we also include a performance ranking and the success rate (ρ).

The performance ranking is calculated as follows. All methods solve every instance in the set. Then, for each instance, the methods are sorted based on the profit of their solutions. The best one receives a ranking of 1, the second one a ranking of 2, and so on. In the case of ties, the ranking is the average of tied positions. This metric is helpful to identify ties that indicate a similar performance of two or more methods. Conversely, the success rate is calculated as the percentage of instances in a set where a particular method reaches or surpasses a threshold. In this investigation, such a threshold is given per instance and corresponds to the best result among those achieved by the heuristics, i.e., the best profit we can get from using each heuristic individually. For the sake of clarity, we present the success rate as a vector of two components: one where the threshold is achieved and one where it is surpassed. Bear in mind that the second component in the vector can only be achieved by mixing heuristics throughout the solving process, and so it only applies to HHs.

4. Experiments and Results

We organized our experiments in three stages: preliminary, exploratory, and confirmatory experiments. For the sake of clarity, these stages are further divided into categories. Figure 2 provides an overview of our experimental methodology. In the following lines, we carefully describe each stage, the corresponding experiments, and the results we obtained.

4.1. Preliminary Experiments

The preliminary experiments were designed first to justify the need for an intelligent way to combine the heuristics (the HHs) and, second, to determine a suitable set of parameters for running the EA to produce such HHs.

4.1.1. Preliminary Experiment I

The first experiment conducted in this work aims to prove that we need an intelligent approach to define the sequences of heuristics. In other words, we justify that it is unlikely for a random sequence of heuristics to achieve competent results. For this purpose, we generated 30 random HHs. The length of these HHs was set to 16 heuristics for empirical reasons. As in this experiment, we were only interested in the behavior of randomly generated HHs, the EA was not used to improve the initial HHs. Then, we used the 30 randomly generated HHs to solve set α_2, and we compared their results against the ones obtained by the available heuristics (see Section 3 for more details). For the sake of clarity, our comparisons include the results of three representative HHs from the perspective of total profit on set α_2: the best, the median, and the worst hyper-heuristics.

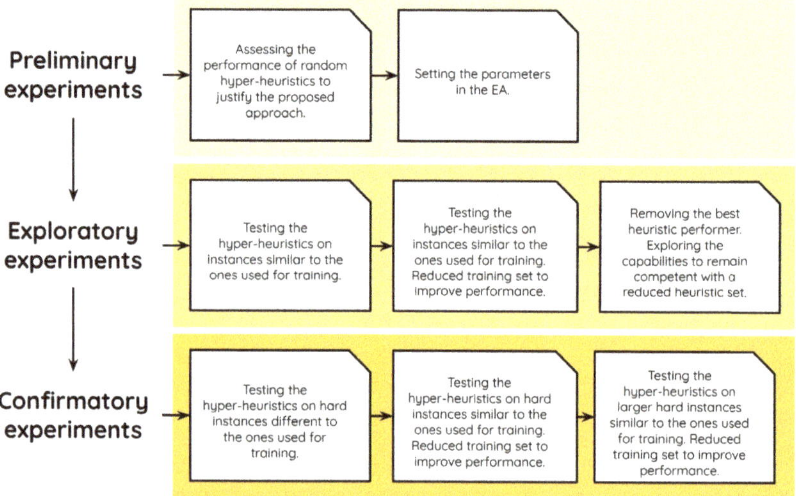

Figure 2. Three-stage methodology followed in this work.

One way to analyze the performance of the methods relies on ranking the resulting data. Table 2 shows the ranks obtained by the four heuristics, as well as the three representative HHs, when solving set α_2. As one may observe, in most of the cases, the heuristics obtain the best result in isolation (ranks close to 1), as it was expected as set α_2 is balanced. Although the best HH never occupies the last positions in the rank, the median performance hints that the good behavior of `Best-HH` is more likely to be due to chance. Furthermore, by randomly generating the sequences of heuristics, we risk producing something as bad as `Worst-HH`, where it replicated the worst heuristic for this set, i.e., `Def`.

Table 2 also shows the total profit obtained for each method on set α_2, which serves as evidence that it is indeed possible to produce one good HH at random, such that it outperforms several heuristics. However, this seems unlikely to happen by chance since at least half the HHs perform worse than `MinW`.

Regarding the success rate, the results confirm that generating sequences of heuristics at random is not a good idea. The success rate of the best randomly generated HH was $(0.0500, 0.0025)$, which means that only in 5% of the instances in set α_2 this HH produced a solution as competent as the best one among the four heuristics, while in 0.25% of the instances for the same set, the HH improved upon this result.

4.1.2. Preliminary Experiment II

Before moving further in this investigation, we needed to fix the number of iterations for the EA. Therefore, we generated 30 HHs by running the EA for 1000 iterations each time. In all the cases, the initial HH contained sequences of 12 heuristics, and we used α_1 as the training set. This time, the cardinality of the HHs was reduced by $1/4$ concerning the previous experiment since the mutation operators allow shortening and extending the sequences. Then, we no longer need long initial sequences as with the first preliminary experiment. For each HH, we recorded the Stagnation Point (SP). We define SP as the iteration at which the best solution was first encountered and for which the profit showed no improvement in subsequent iterations. Table 3 shows the stagnation points of the 30 runs of the EA.

Table 2. Ranks and total profit obtained by the four heuristics and the best, median, and worst randomly generated hyper-heuristics when tested on set α_2. The best result, measured in terms of total profit, is highlighted in bold.

	Heuristics				Hyper-Heuristics		
Rank	Def	MaxP	MaxPW	MinW	Best-HH	Median-HH	Worst-HH
1.0	1	100	102	78	1	0	0
1.5	80	0	9	22	12	20	79
2.0	4	0	111	0	107	1	4
2.5	6	0	59	0	64	5	6
3.0	0	13	20	35	83	65	1
3.5	9	1	82	4	99	21	10
4.0	0	27	15	90	29	80	0
4.5	2	0	2	43	3	46	2
5.0	0	165	0	28	1	103	0
5.5	17	0	0	34	0	34	17
6.0	0	12	0	32	1	6	0
6.5	281	0	0	16	0	16	281
7.0	0	82	0	18	0	3	0
Profit	241,081	343,872	**467,145**	376,148	436,393	361,284	241,025

Table 3. Stagnation points for the first 30 runs of the EA-based hyper-heuristic model.

Run	SP	Run	SP	Run	SP
1	162	11	46	21	675
2	127	12	51	22	41
3	743	13	114	23	30
4	39	14	38	24	47
5	682	15	18	25	40
6	38	16	317	26	28
7	68	17	56	27	152
8	21	18	146	28	78
9	405	19	176	29	37
10	64	20	30	30	34

We used these stagnation points to estimate the maximum number of iterations for running the EA. The average stagnation point among the 30 runs was 150.1, so we rounded it down to 150 iterations. Fifty additional iterations were added just as a precaution to minimize the risk of not reaching a good result. Thus, we set the maximum number of iterations to 200 for the rest of the experiments.

4.2. Exploratory Experiments

The exploratory experiments comprise a series of tests that cover general aspects of the proposed model, particularly those regarding how it copes with single heuristics on the balanced instance sets (sets α_1 and α_2).

4.2.1. Exploratory Experiment I

In this experiment, the rationale was to test the performance of HHs on similar instances to those they were trained on. Therefore, we produced 30 HHs with an initial cardinality of 12 heuristics each, as in preliminary experiment 2 (Section 4.1.2). Each HH was trained using set α_1 for 200 iterations. Afterward, we tested the resulting HHs on set α_2 and we compared the data against those obtained by heuristics applied in isolation.

Table 4 presents the ranking of the four heuristics as well as the best, median, and worst HHs produced in this experiment and tested on set α_2. This table also shows the total profit obtained by each of these methods on the same set. Based on the results obtained (both on ranks and total profit), we observe that the process is forcing the HHs to replicate the behavior of the best performing heuristic for these types of instances (MaxPW). This also means that the HHs are, most of the time, ignoring the remaining heuristics. Although this may seem like a good choice as MaxPW is the best individual performer, the HHs are sacrificing the opportunity to improve their overall performance and outperform the best individual heuristic.

Table 4. Ranks and total profit obtained by the four heuristics and the best, median, and worst hyper-heuristics trained on set α_1 and tested on set α_2. The best result, measured in terms of total profit, is highlighted in bold.

	Heuristics				Hyper-Heuristics		
Rank	Def	MaxP	MaxPW	MinW	Best-HH	Median-HH	Worst-HH
1.0	85	100	0	87	0	7	0
1.5	0	0	6	8	6	0	8
2.0	1	1	5	4	5	16	1
2.5	0	0	100	1	100	99	100
3.0	2	0	3	0	3	10	5
3.5	1	0	264	1	264	258	258
4.0	0	0	16	1	16	4	25
4.5	0	0	6	0	6	1	1
5.0	11	21	0	78	0	5	2
5.5	0	1	0	1	0	0	0
6.0	18	195	0	183	0	0	0
7.0	282	82	0	36	0	0	0
Profit	241,081	343,872	**467,145**	376,148	**467,145**	467,103	466,711

4.2.2. Exploratory Experiment 2

In the previous experiment, the EA forced the HHs into replicating the behavior of one heuristic, MaxPW. In this experiment, we try to overcome this situation by reducing the number of instances in the training set. Then, for this experiment, 30 new HHs were produced, but this time, only 60% of the instances in set α_1 were used for training purposes. These 60% instances were randomly selected once and used for producing the 30 HHs. As in the previous experiment, the maximum number of iterations for the EA was set to 200 and all the HHs were initialized by using a random sequence of 12 heuristics. We used the 30 HHs to solve set α_2 and summarized the results through the rankings and total profit (Table 5).

Based on the results shown in Table 5, we observe a small improvement in Best-HH with respect to MaxPW. Although this supports our initial idea that we can improve the results obtained by the heuristics with a hyper-heuristic, the improvement is rather small and insignificant in practice. Furthermore, the success rate of Best-HH is $(0.285, 0.0)$, which is exactly the same as MaxPW.

For a more in-depth analysis about the performance of the best HH in this experiment, we plotted the performance of Best-HH, as well as of the best and worst heuristic for each particular instance in set α_2. For clarity, the results are sorted (from larger to smaller) on the profit obtained by Best-HH on each instance in the set. Figure 3 depicts the profit of each method across all instances. As we can observe, there are many cases where Best-HH obtains a smaller profit than the best heuristic for each particular instance.

Table 5. Ranks and total profit obtained by the four heuristics and the best, median, and worst hyper-heuristics trained on 60% of set α_1 and tested on set α_2. The best result, measured in terms of total profit, is highlighted in bold.

	Heuristics				Hyper-Heuristics		
Rank	Def	MaxP	MaxPW	MinW	Best-HH	Median-HH	Worst-HH
1.0	86	100	0	91	0	0	0
1.5	0	0	5	4	0	5	4
2.0	0	1	6	5	11	6	5
2.5	0	0	100	0	100	100	100
3.0	2	0	4	0	9	4	2
3.5	1	0	274	1	274	274	274
4.0	0	0	5	1	5	5	12
4.5	0	0	6	0	1	6	1
5.0	11	21	0	78	0	0	2
5.5	0	1	0	1	0	0	0
6.0	18	195	0	183	0	0	0
7.0	282	82	0	36	0	0	0
Profit	241,081	343,872	467,145	376,148	**467,182**	467,145	466,711

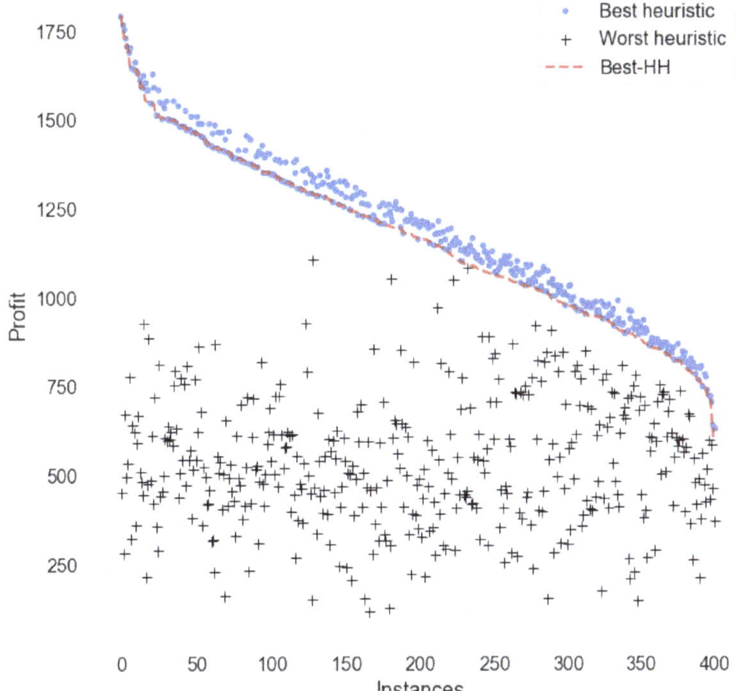

Figure 3. Profit per instance obtained by the best hyper-heuristic trained on 60% of set α_1 and the best and worst heuristic per instance on set α_2.

4.2.3. Exploratory Experiment III

So far, we have observed that, even though it is possible to overcome the best individual performer for each instance in some specific cases, oftentimes the HHs tend to replicate the behavior of the overall best heuristic (MaxPW for sets α_1 and α_2). Although this is not a bad result—the model produces very competent HHs, it is difficult to justify the additional time devoted to producing such HHs given the small gains with respect to MaxPW. For this reason, in this experiment we wanted to explore the case where the HHs can only choose among Def, MaxP, and MinW (all the available heuristics but MaxPW) and evaluate if the HHs produced may still be considered competent. As in the previous experiments, we generated 30 HHs by training on set α_1 for 200 iterations each, and testing on set α_2. In all the cases, the HHs have an initial cardinality of 12 heuristics.

A comparison between the total profits of Best-HH and MaxPW (Table 6) reveals that a significant efficiency was lost due to the removal of MaxPW from the pool of heuristics (Figure 4). However, all three representative HHs (best, median, and worst) obtained the second rank in terms of total profit. The profit of Best-HH shows an improvement of nearly 6% and over 64% with respect to the MinW and Def heuristics, respectively. Therefore, the model can learn in harsh conditions and thus obtains promising results regardless of whether it was given a pool of heuristics with limited quality.

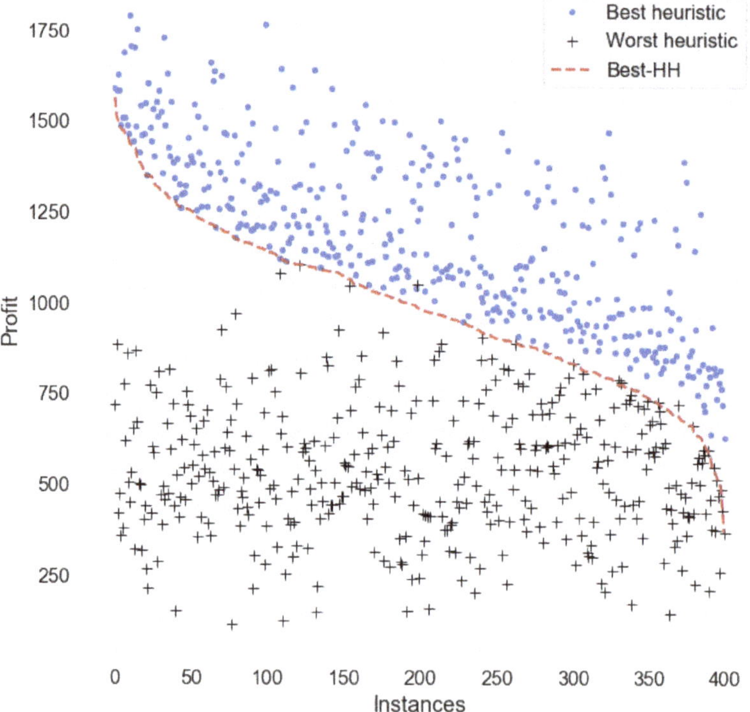

Figure 4. Profit per instance obtained by the best hyper-heuristic trained on set α_1 (removing MaxPW from the pool of available heuristics) and the best and worst heuristic per instance on set α_2 (including MaxPW in the comparison).

4.3. Confirmatory Experiments

Seeking to test the generalization properties of the proposed HH model, we conducted three additional experiments that now incorporate problem instances taken from the literature. These experiments were conducted in a similar way to the exploratory ones: each one involves the generation of 30 HHs with an initial cardinality of 12, which were trained for 200 iterations. However, this time we tried some combinations of the instance sets for training and testing. The rationale behind these experiments is to observe how well the HHs adapt to changes in the properties of the instances being solved with respect to the instances used for producing such HHs.

Table 6. Ranks and total profit obtained by the four heuristics and the best, median, and worst hyper-heuristics trained on set α_1 and tested on set α_2, but MaxPW is not available for the hyper-heuristics. The best result, measured in terms of total profit, is highlighted in bold.

	Heuristics				Hyper-Heuristics		
Rank	Def	MaxP	MaxPW	MinW	Best-HH	Median-HH	Worst-HH
1.0	83	99	111	96	0	0	0
1.5	3	0	2	2	1	2	2
2.0	13	2	219	1	49	39	8
2.5	0	0	5	3	19	32	31
3.0	1	14	30	19	82	61	37
3.5	0	0	5	1	49	87	106
4.0	1	4	14	16	70	51	59
4.5	0	0	1	3	33	67	84
5.0	0	26	13	91	53	33	42
5.5	0	1	0	12	17	18	22
6.0	17	176	0	129	18	6	5
6.5	0	0	0	6	7	2	3
7.0	282	78	0	21	2	2	1
Profit	241,081	343,872	**467,145**	376,148	397,373	395,586	391,940

4.3.1. Confirmatory Experiment I

So far, we have evaluated the performance of HHs on instances similar to the ones used for training, under various scenarios. In the first confirmatory experiment, we use all the instances from set α_1 to train the HHs, but test their performance on set β_1. Let us recall that set β_1 contains 600 hard instances proposed by Pisinger [45] and feature 20 items and different capacities. The relevant issue regarding set β_1 is that such instances are considered hard to solve.

Table 7 shows that MaxPW is, once again, the best individual performer in this experiment. Although all the produced HHs perform similarly (cf. Worst-HH), they are unable to improve upon the results obtained by the best heuristic. Nonetheless, all the HHs produced obtained a total profit larger than those obtained by the remaining heuristics.

Table 7. Ranks and total profit obtained by the four heuristics and the best, median, and worst hyper-heuristics trained on set α_1 and tested on set β_1. The best result, measured in terms of total profit, is highlighted in bold.

	Heuristics				Hyper-Heuristics		
Rank	Def	MaxP	MaxPW	MinW	Best-HH	Median-HH	Worst-HH
1.0	24	116	32	10	42	39	31
1.5	16	20	71	11	60	39	25
2.0	32	41	111	79	124	57	33
2.5	14	19	80	27	56	54	54
3.0	54	47	63	38	59	84	77
3.5	18	4	62	16	48	47	43
4.0	69	33	47	35	53	98	79
4.5	30	3	42	14	32	44	47
5.0	63	31	26	48	46	73	94
5.5	22	4	31	17	21	13	24
6.0	190	43	20	64	36	42	75
6.5	15	6	13	14	5	2	3
7.0	53	233	2	227	18	8	15
Profit	3,804,271	3,724,588	**4,039,708**	3,867,345	4,031,981	3,958,061	3,905,734

Even if the HHs were incapable of overcoming MaxPW in this experiment, it is interesting to see how close their performance is, especially as these HHs were trained with instances of a different type than those used for testing.

4.3.2. Confirmatory Experiment II

In the previous experiment, we evaluated the performance of HHs trained on balanced sets of instances when tested on hard instances, and we observed a limited capacity to deal with such instances. In this experiment, we show that this behaviour is not due to the hardness of the instances, but because of the discrepancy between the instances used for training and the ones used for testing. For this reason, this experiment is twofold: (1) we test how HHs behave when trained and tested on hard instances (set β_1) and (2) we try to reduce the effect of MaxPW in the resulting HHs by reducing the number of instances in the training set. Then, we produced 30 new HHs using only 60% of the instances in set β_1 for training purposes. As in previous experiments, the 60% instances were randomly selected once and used for producing the 30 HHs. For consistency, the maximum number of iterations for the EA was set to 200 and all the HHs were initialized to a cardinality of 12. The HHs produced in this experiment were evaluated on the remaining 40% of the instances in set β_1. The results for rankings and total profit derived from this experiment are summarized in Table 8.

Table 8. Ranks and total profit obtained by the four heuristics and the best, median, and worst hyper-heuristics trained on 60% of set β_1 and tested on the remaining 40% of set β_1. The best result, measured in terms of total profit, is highlighted in bold.

	Heuristics				Hyper-Heuristics		
Rank	Def	MaxP	MaxPW	MinW	Best-HH	Median-HH	Worst-HH
1.0	16	54	3	3	12	4	2
1.5	1	0	6	5	13	3	16
2.0	14	8	10	9	29	28	20
2.5	2	5	47	3	51	59	59
3.0	14	16	54	30	67	64	56
3.5	3	3	20	0	19	25	22
4.0	19	13	24	12	16	21	13
4.5	4	1	29	11	18	20	21
5.0	30	19	10	19	10	8	11
5.5	9	3	21	13	4	6	6
6.0	101	16	9	28	1	0	10
6.5	6	4	6	10	0	2	2
7.0	21	98	1	97	0	0	2
Profit	1,584,007	1,540,100	1,673,387	1,600,015	**1,681,403**	1,681,127	1,678,375

This experiment confirms the importance of the instances used for training and their similarity with those solved during the test process. This time, the performance of most of the hyper-heuristics produced using 60% of set β_1 when tested on the remaining 40% of set β_1 improve on MaxPW (the best heuristic for this set). In fact, the performance of Worst-HH is practically the same as MaxPW (Figure 5). Although the results suggest that an improvement in the total profit can be obtained by using the proposed hyper-heuristic approach, the gain in profit derived from using Best-HH instead of MaxPW is rather small ($<0.5\%$). However, this result should not be interpreted as an indication that Best-HH is replicating the behavior of MaxPW. In fact, their overall profit is similar (with a small gain for Best-HH but their specific performance is not. As depicted in Figure 6, when the ranks are analyzed, we observe that Best-HH obtains better results in more cases than the best individual heuristic, MaxPW.

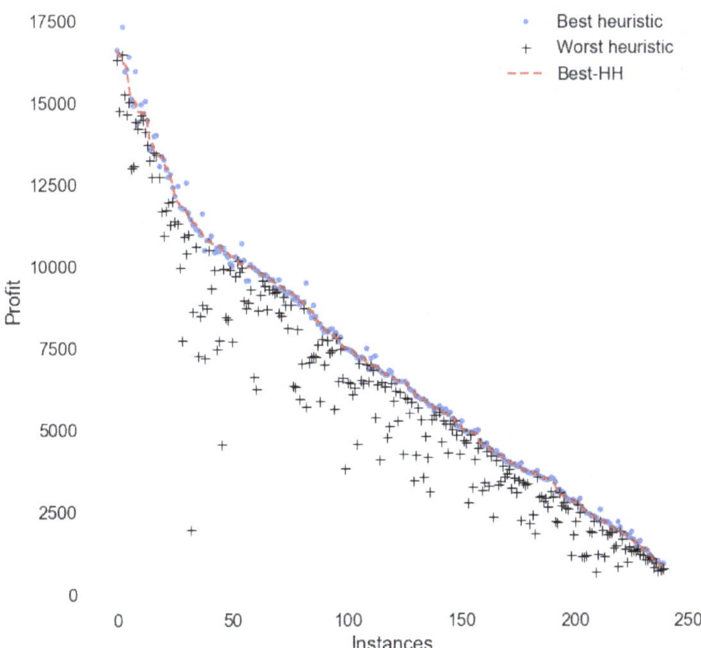

Figure 5. Profit per instance obtained by the best hyper-heuristic trained on 60% of set β_1, and the best and worst heuristic per instance on the remaining 40% of set β_1.

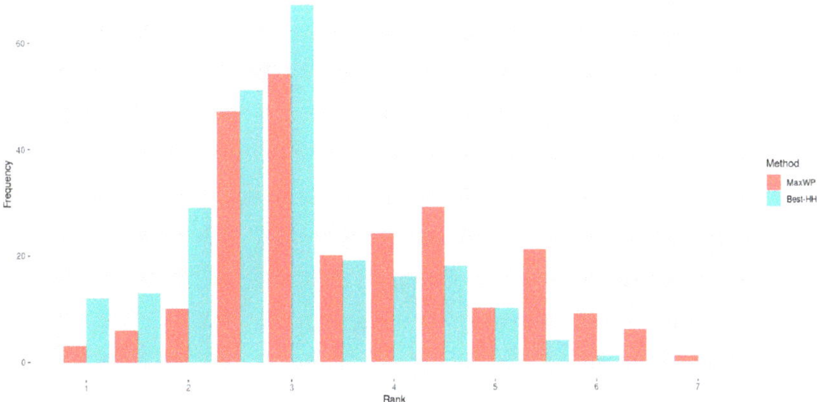

Figure 6. Frequency of the ranks obtained by MaxPW and Best-HH (trained on 60% of set β_1) the remaining 40% of set β_1.

4.3.3. Confirmatory Experiment III

In this final experiment, we extend our analysis to hard and larger instances. This time, we produced 30 HHs by using 60% of set β_2 and tested them on the remaining 40% of the same set. With this experiment, we validate that the learning method is quite stable as it still produces competent hyper-heuristics. Despite the fact that the set comprised instances with different features, all three cases beat the best operator (MaxPW) in isolation. Additionally, setting a good training set seems to impact the efficiency of the HH model.

Training done on α_1 seemed to negatively affect the results, as seen on exploratory experiment I and confirmatory experiment I. It is important to note that sets α_1 and α_2 are balanced and synthetically produced. Thus, 25% of the problem instances are designed to maximize the performance of a single low-level heuristic. This pattern is repeated for all the heuristics in this set. Conversely, hard problem instances from sets β_1 and β_2 are randomly sampled (without replacement) using three different seeds. This training scheme is more representative of real-life applications, where often no balanced or ideal conditions are met.

The model's behavior is similar to what we observed in previous cases. The hyper-heuristics trained on hard and larger instances are still competitive regarding the single heuristics applied in isolation. This time, the improvement on MaxPW is smaller than in previous cases (<0.01%). Table 9 shows that, as in previous cases, there is a difference in the behavior of the heuristics and HHs in terms of ranks, but it does not necessarily represent a significant difference in terms of total profit. However, when we analyze the success rate, we observe that Best-HH obtains promising results, $(0.6542, 0.0958)$. This indicates that in little more of 65% of the instances in the 40% of set β_2 used for testing, Best-HH was equal in performance to the best out of the four heuristics. Besides, in almost 10% of the instances for the same set, Best-HH improved upon this result.

Table 9. Ranks and total profit obtained by the four heuristics and the best, median, and worst hyper-heuristics trained on 60% of set β_2 and tested on the remaining 40% of set β_2. The best result, measured in terms of total profit, is highlighted in bold.

	Heuristics				Hyper-Heuristics		
Rank	Def	MaxP	MaxPW	MinW	Best-HH	Median-HH	Worst-HH
1.0	10	20	0	15	10	4	15
1.5	1	1	4	2	8	6	2
2.0	10	12	14	2	23	19	7
2.5	0	1	81	2	80	82	80
3.0	14	25	62	33	62	71	63
3.5	1	2	31	3	29	30	26
4.0	3	5	9	9	8	6	16
4.5	0	0	17	6	10	11	12
5.0	25	27	4	44	2	3	4
5.5	3	2	15	8	7	8	9
6.0	140	34	2	27	1	0	5
6.5	1	1	1	0	0	0	1
7.0	32	110	0	89	0	0	0
Profit	3,752,072	3,631,322	4,046,715	3,930,092	**4,052,501**	4,052,039	4,045,794

4.4. Discussion

We would like to discuss the rationale behind the proposed model and why it may be useful for other problem domains besides the one studied in this document.

As with other packing problems, the iterative nature of KP makes it a great candidate for a hyper-heuristic approach. The inherent mapping of problem states to decisions (or, in this case, packing heuristics) can lead to an optimal item selection by overfitting if the relationship were to be searched exhaustively. A generalization of this item selection is the basis of the rationale behind this approach. Furthermore, the advantages of mixing heuristics have previously been discussed in detail for various optimization and search problems [27,58].

The evidence obtained from the experiments confirms that it is possible to produce a sequence of heuristics that provides an acceptable performance when tested on a set of instances. Allow us to explain this in more detail. Let us assume that a hyper-heuristic HH_1 is to be produced for solving only one KP instance with n items, KP_1. If the cardinality of the hyper-heuristic is equal to the number of items to pack ($\varpi = n$), then HH_1 can decide which heuristic to use for packing each item. Among all the possible HHs that could be produced, there is one where all the decisions are correct, HH_1^*, which represents the optimal sequence to pack the items in KP_1, given the available heuristics. Now, let us produce a new hyper-heuristic HH_2^*, the best sequence of heuristics for solving a second KP instance, KP_2, also containing n items (the simplest case for the analysis). There is no guarantee that the previous hyper-heuristic, HH_1^*, will also represent the optimal sequence of heuristics for solving KP_2. If we keep the idea of having one individual decision per item, only a few of these decisions will be the same for both sequences, HH_1^* and HH_2^*. In order to merge the two sequences, some errors must be accepted. The task of the evolutionary framework is to decide which errors to accept so that the performance is the best among all the instances in the training set. Overall, the model is not looking for the best sequence of heuristics for each particular instance but the best sequence to solve, acceptably, all the instances in the training set.

5. Conclusions and Future Work

In this work, we analyzed the efficiency of a selection HH which does not depend on problem characterization. The analysis showed that the learning mechanism deals very well on all instances despite its simple parameters. For small instances like the ones in sets α_1 and α_2, the MaxPW heuristic seemed like the most suitable heuristic in isolation. The instances used for training were generated with one packing heuristic in mind. Instead, instances in the literature are considered harder and represent a challenge for a single heuristic in isolation. It is also important to note that the instance sets may be considered small, so the learning process was not computationally expensive. For larger datasets with more instances to solve, a HH could perform better if one has adequate resources for the learning process.

Our work proves the feasibility of a feature-independent hyper-heuristic approach powered by an evolutionary algorithm. The results confirm our idea that it is possible to generate generic sequences of heuristics that perform better than individual heuristics for specific sets of instances. Our results also demonstrate that the similarity between the training and testing instances influences the model's performance. In other words, the model can generalize to unseen types of instances, but the ideal scenario would be to train the hyper-heuristics on instances with similar features to the ones to be solved when training is over. At this point, we consider it fair to mention that we are aware of the diversity of the optimization problems. We have selected the KP because it was convenient for our study, as we can easily develop and test hyper-heuristics for this particular problem. However, many other exciting problems could have also been considered in this regard. Our interest was to propose a new hyper-heuristic method to deal with instances that are hard to solve by exact methods.

Many considerations could be taken into account to improve the analysis. Fixing the size of HHs and setting a single heuristic per gene in the model may impact the frequency analysis. Adding more mutation operators and tweaking their probability distribution is also something to consider for future work. More importantly, extending the amount of packing heuristics to choose from, though increasing the heuristic search space, may display more explicit differences between heuristic sequences. Once again, we would like to emphasize that the selection HH proposed did not deal with any problem characterization or feature analysis. Adding problem characterization may improve the performance of the learning process even more at the expense of some additional computational effort.

Although we have not tested the proposed model on other NP-hard optimization problems, we are confident that the model can work correctly on similar problems such as packing and scheduling. Based on our experience, those problems show similar properties that allow hyper-heuristics to grasp the structure of the instances and decide which heuristic to apply at certain times of the solving process. Therefore, the proposed model should perform properly. Nonetheless, testing our approach on other challenging problems is, undoubtedly, a path for future work.

Author Contributions: Conceptualization, X.S.-D. and J.C.O.-B.; methodology, X.S.-D., I.A. and J.M.C.-D.; software, X.S.-D. and J.C.O.-B.; validation, I.A. and J.M.C.-D.; formal analysis, X.S.-D.; investigation, X.S.-D., H.T.-M. and S.E.C.-P.; resources, J.C.O.-B.; data curation, X.S.-D.; writing—original draft preparation, X.S.-D.; writing—review and editing, J.C.O.-B., I.A. and J.M.C.-D.; visualization, X.S.-D., I.A. and J.M.C.-D.; supervision, H.T.-M. and S.E.C.-P.; project administration, J.C.O.-B.; funding acquisition, I.A. All authors have read and agreed to the published version of the manuscript.

Funding: This research was partially funded by the research group with strategic focus in intelligent systems at ITESM, grant NUA A00834075, and CONACyT Basic Science projects with grant number 287479 and fellowship 2021-000001-01NACF-00604.

Institutional Review Board Statement: Not applicable.

Informed Consent Statement: Not applicable.

Conflicts of Interest: The authors declare no conflict of interest.

References

1. Sánchez, M.; Cruz-Duarte, J.M.; Ortiz-Bayliss, J.C.; Ceballos, H.; Terashima-Marín, H.; Amaya, I. A Systematic Review of Hyper-heuristics on Combinatorial Optimization Problems. *IEEE Access* **2020**, 1–28. [CrossRef]
2. Bai, R.; Burke, E.K.; Gendreau, M.; Kendall, G.; McCollum, B. Memory Length in Hyper-heuristics: An Empirical Study. In Proceedings of the IEEE Symposium on Computational Intelligence in Scheduling, SCIS '07, Honolulu, HI, USA, 2–4 April 2007; pp. 173–178. [CrossRef]
3. Burke, E.K.; Gendreau, M.; Hyde, M.; Kendall, G.; Ochoa, G.; Özcan, E.; Qu, R. Hyper-heuristics: A survey of the state of the art. *J. Oper. Res. Soc.* **2013**, *64*, 1695–1724. [CrossRef]
4. Pillay, N.; Qu, R. *Hyper-Heuristics: Theory and Applications*; Natural Computing Series; Springer International Publishing: Cham, Switzerland, 2018; doi:10.1007/978-3-319-96514-7. [CrossRef]
5. Poli, R.; Graff, M. There is a free lunch for hyper-heuristics, genetic programming and computer scientists. In Proceedings of the 12th European Conference on Genetic Programming (EuroGP 2009), Tübingen, Germany, 15–17 April 2009; Springer: Berlin/Heidelberg, Germany, 2009; pp. 195–207._17. [CrossRef]
6. Özcan, E.; Bilgin, B.; Korkmaz, E.E. A Comprehensive Analysis of Hyper-heuristics. *Intell. Data Anal.* **2008**, *12*, 3–23. [CrossRef]
7. Hart, E.; Sim, K. A Hyper-Heuristic Ensemble Method for Static Job-Shop Scheduling. *Evol. Comput.* **2016**, *24*, 609–635. [CrossRef] [PubMed]
8. Hyde, M. A Genetic Programming Hyper-Heuristic Approach to Automated Packing. Ph.D. Thesis, University of Nottingham, Nottingham, UK, 2010.
9. Drake, J.H.; Hyde, M.; Ibrahim, K.; Özcan, E. A Genetic Programming Hyper-Heuristic for the Multidimensional Knapsack Problem. In Proceedings of the 11th IEEE International Conference on Cybernetic Intelligent Systems, Limerick, Ireland, 23–24 August 2012.
10. Lourenço, N.; Pereira, F.B.; Costa, E. The Importance of the Learning Conditions in Hyper-heuristics. In Proceedings of the 15th Annual Conference on Genetic and Evolutionary Computation, GECCO '13, Amsterdam, The Netherlands, 6–10 July 2013; ACM: New York, NY, USA, 2013; pp. 1525–1532. [CrossRef]
11. Falkenauer, E. A hybrid grouping genetic algorithm for bin packing. *J. Heuristics* **1996**, *2*, 5–30. [CrossRef]
12. Ortiz-Bayliss, J.C.; Moreno-Scott, J.H.; Terashima-Marín, H. Automatic Generation of Heuristics for Constraint Satisfaction Problems. In *Nature Inspired Cooperative Strategies for Optimization (NICSO 2013)*; Studies in Computational Intelligence; Springer: Cham, Switzerland, 2013; Volume 512, pp. 315–327._24. [CrossRef]
13. Hart, E.; Sim, K. On the Life-Long Learning Capabilities of a NELLI*: A Hyper-Heuristic Optimisation System. In Proceedings of the International Conference on Parallel Problem Solving from Nature—PPSN XIII, Ljubljana, Slovenia, 13–17 September 2014; Volume 8672, pp. 282–291.
14. Sim, K.; Hart, E. An Improved Immune Inspired Hyper-heuristic for Combinatorial Optimisation Problems. In Proceedings of the 2014 Annual Conference on Genetic and Evolutionary Computation, GECCO '14, Vancouver, BC, Canada, 12–16 July 2014; ACM: New York, NY, USA, 2014; pp. 121–128. [CrossRef]
15. Amaya, I.; Ortiz-Bayliss, J.C.; Conant-Pablos, S.; Terashima-Marin, H. Hyper-heuristics Reversed: Learning to Combine Solvers by Evolving Instances. In Proceedings of the 2019 IEEE Congress on Evolutionary Computation (CEC), Wellington, New Zealand, 10–13 June 2019; pp. 1790–1797. [CrossRef]
16. Hart, E.; Sim, K. On Constructing Ensembles for Combinatorial Optimisation. In Proceedings of the Genetic and Evolutionary Computation Conference Companion, GECCO '17, Berlin, Germany, 15–19 July 2017; ACM: New York, NY, USA, 2017; pp. 5–6. [CrossRef]
17. Burke, E.K.; Hyde, M.R.; Kendall, G.; Woodward, J. Automating the Packing Heuristic Design Process with Genetic Programming. *Evol. Comput.* **2012**, *20*, 63–89._a_00044. [CrossRef]

18. Drake, J.H.; Özcan, E.; Burke, E.K. Modified Choice Function Heuristic Selection for the Multidimensional Knapsack Problem. In *Genetic and Evolutionary Computing*; Sun, H., Yang, C.Y., Lin, C.W., Pan, J.S., Snasel, V., Abraham, A., Eds.; Springer International Publishing: Cham, Switzerland, 2015; pp. 225–234.

19. Drake, J.H.; Özcan, E.; Burke, E.K. A Case Study of Controlling Crossover in a Selection Hyper-heuristic Framework Using the Multidimensional Knapsack Problem. *Evol. Comput.* **2016**, *24*, 113–141._a_00145. [CrossRef]

20. Duhart, B.; Camarena, F.; Ortiz-Bayliss, J.C.; Amaya, I.; Terashima-Marín, H. An Experimental Study on Ant Colony Optimization Hyper-Heuristics for Solving the Knapsack Problem. In *Pattern Recognition*; Martínez-Trinidad, J.F., Carrasco-Ochoa, J.A., Olvera-López, J.A., Sarkar, S., Eds.; Springer International Publishing: Cham, Switzerland, 2018; pp. 62–71.

21. Gómez-Herrera, F.; Ramirez-Valenzuela, R.A.; Ortiz-Bayliss, J.C.; Amaya, I.; Terashima-Marín, H. A Quartile-Based Hyper-heuristic for Solving the 0/1 Knapsack Problem. In *Advances in Soft Computing*; Castro, F., Miranda-Jiménez, S., González-Mendoza, M., Eds.; Springer International Publishing: Cham, Switzerland, 2018; pp. 118–128.

22. Drake, J.H.; Kheiri, A.; Özcan, E.; Burke, E.K. Recent advances in selection hyper-heuristics. *Eur. J. Oper. Res.* **2020**, *285*, 405–428. [CrossRef]

23. Wilbaut, C.; Hanafi, S.; Salhi, S. A survey of effective heuristics and their application to a variety of knapsack problems. *IMA J. Manag. Math.* **2008**, *19*, 227. [CrossRef]

24. Rice, J.R. The Algorithm Selection Problem. *Adv. Comput.* **1976**, *15*, 65–118. [CrossRef]

25. Garza-Santisteban, F.; Sanchez-Pamanes, R.; Puente-Rodriguez, L.A.; Amaya, I.; Ortiz-Bayliss, J.C.; Conant-Pablos, S.; Terashima-Marin, H. A Simulated Annealing Hyper-heuristic for Job Shop Scheduling Problems. In Proceedings of the 2019 IEEE Congress on Evolutionary Computation (CEC), Wellington, New Zealand, 10–13 June 2019; pp. 57–64. [CrossRef]

26. Alanazi, F.; Lehre, P.K. Runtime analysis of selection hyper-heuristics with classical learning mechanisms. In Proceedings of the 2014 IEEE Congress on Evolutionary Computation (CEC), Beijing, China, 6–11 July 2014; pp. 2515–2523. [CrossRef]

27. Lehre, P.K.; Özcan, E. A Runtime Analysis of Simple Hyper-heuristics: To Mix or Not to Mix Operators. In Proceedings of the Twelfth Workshop on Foundations of Genetic Algorithms XII, FOGA XII '13, Adelaide, Australia, 16–20 January 2013; ACM: New York, NY, USA, 2013; pp. 97–104. [CrossRef]

28. Ortiz-Bayliss, J.C.; Terashima-Marín, H.; Özcan, E.; Parkes, A.J.; Conant-Pablos, S.E. Exploring heuristic interactions in constraint satisfaction problems: A closer look at the hyper-heuristic space. In Proceedings of the 2013 IEEE Congress on Evolutionary Computation, Cancun, Mexico, 20–23 June 2013; pp. 3307–3314. [CrossRef]

29. Amaya, I.; Ortiz-Bayliss, J.C.; Rosales-Pérez, A.; Gutiérrez-Rodríguez, A.E.; Conant-Pablos, S.E.; Terashima-Marín, H.; Coello, C.A. Enhancing Selection Hyper-Heuristics via Feature Transformations. *IEEE Comput. Intell. Mag.* **2018**, *13*, 30–41. [CrossRef]

30. Sánchez-Díaz, X.F.C.; Ortiz-Bayliss, J.C.; Amaya, I.; Cruz-Duarte, J.M.; Conant-Pablos, S.E.; Terashima-Marín, H. A Preliminary Study on Feature-independent Hyper-heuristics for the 0/1 Knapsack Problem. In Proceedings of the 2020 IEEE Congress on Evolutionary Computation (CEC), Glasgow, UK, 19–24 July 2020; pp. 1–8. [CrossRef]

31. Dudzinski, K.; Walukiewicz, S. Exact methods for the knapsack problem and its generalizations. *Eur. J. Oper. Res.* **1987**, *28*, 3–21. [CrossRef]

32. Martello, S.; Pisinger, D.; Toth, P. New trends in exact algorithms for the 0–1 knapsack problem. *Eur. J. Oper. Res.* **2000**, *123*, 325–332. [CrossRef]

33. Lawler, E.L. Fast Approximation Algorithms for Knapsack Problems. *Math. Oper. Res.* **1979**, *4*, 339–356. [CrossRef]

34. Lienland, B.; Zeng, L. A Review and Comparison of Genetic Algorithms for the 0–1 Multidimensional Knapsack Problem. *Int. J. Oper. Res. Inf. Syst.* **2015**, *6*, 21–31. [CrossRef]

35. Hembecker, F.; Lopes, H.; Godoy, W., Jr. Particle Swarm Optimization for the Multidimensional Knapsack Problem. In Proceedings of the International Conference on Adaptive and Natural Computing Algorithms, Warsaw, Poland, 11–14 April 2007; Volume 4331, pp. 358–365._40. [CrossRef]

36. Gómez, N.; López, L.; Albert, A. Multidimensional knapsack problem optimization using a binary particle swarm model with genetic operations. *Soft Comput.* **2018**, *22*, 2567–2582. [CrossRef]

37. Razavi, S.; Sajedi, H. Cognitive discrete gravitational search algorithm for solving 0–1 knapsack problem. *J. Intell. Fuzzy Syst.* **2015**, *29*, 2247–2258. [CrossRef]

38. Patvardhan, C.; Bansal, S.; Srivastav, A. Quantum-Inspired Evolutionary Algorithm for difficult knapsack problems. *Memetic Comput.* **2015**, *7*. [CrossRef]

39. Lv, J.; Wang, X.; Huang, M.; Cheng, H.; Li, F. Solving 0–1 knapsack problem by greedy degree and expectation efficiency. *Appl. Soft Comput. J.* **2016**, *41*, 94–103. [CrossRef]

40. Kulkarni, A.J.; Shabir, H. Solving 0–1 Knapsack Problem using Cohort Intelligence Algorithm. *Int. J. Mach. Learn. Cybern.* **2016**, *7*, 427–441. [CrossRef]

41. Banda, J.; Velasco, J.; Berrones, A. A hybrid heuristic algorithm based on Mean-Field Theory with a Simple Local Search for the Quadratic Knapsack Problem. In Proceedings of the 2017 IEEE Congress on Evolutionary Computation (CEC), San Sebastián, Spain, 5–8 June 2017; pp. 2559–2565. [CrossRef]

42. Martello, S.; Toth, P. *Knapsack Problems: Algorithms and Computer Implementations*; John Wiley & Sons, Inc.: Hoboken, NJ, USA, 1990.

43. Kellerer, H.; Pferschy, U.; Pisinger, D. *Knapsack Problems*; Springer: Berlin, Germany, 2004.

44. Furini, F.; Monaci, M.; Traversi, E. Exact approaches for the knapsack problem with setups. *Comput. Oper. Res.* **2018**, *90*. [CrossRef]

45. Pisinger, D. Where Are the Hard Knapsack Problems? *Comput. Oper. Res.* **2005**, *32*, 2271–2284. [CrossRef]
46. Sun, X.; Sheng, H.; Li, D. An exact algorithm for 0–1 polynomial knapsack problems. *J. Ind. Manag. Optim.* **2007**, *3*, 223–232. [CrossRef]
47. Burke, E.K.; Hyde, M.R.; Kendall, G.; Ochoa, G.; Özcan, E.; Woodward, J.R. A classification of hyper-heuristic approaches: Revisited. In *Handbook of Metaheuristics*; Springer: Cham, Switzerland, 2019; pp. 453–477.
48. Dokeroglu, T.; Sevinc, E.; Kucukyilmaz, T.; Cosar, A. A survey on new generation metaheuristic algorithms. *Comput. Ind. Eng.* **2019**, *137*, 106040. [CrossRef]
49. Vikhar, P.A. Evolutionary algorithms: A critical review and its future prospects. In Proceedings of the 2016 IEEE International Conference on Global Trends in Signal Processing, Information Computing and Communication (ICGTSPICC), Jalgaon, India, 22–24 December 2016; pp. 261–265.
50. Cruz-Duarte, J.M.; Amaya, I.; Ortiz-Bayliss, J.C.; Conant-Pablos, S.E.; Terashima-Marín, H. A Primary Study on Hyper-Heuristics to Customise Metaheuristics for Continuous Optimisation. In Proceedings of the 2020 IEEE Congress on Evolutionary Computation, Glasgow, UK, 19–24 July 2020.
51. Cruz-Duarte, J.M.; Ortiz-Bayliss, J.C.; Amaya, I.; Shi, Y.; Terashima-Marín, H.; Pillay, N. Towards a Generalised Metaheuristic Model for Continuous Optimisation Problems. *Mathematics* **2020**, *8*, 2046. [CrossRef]
52. Salov, V. Notation for Iteration of Functions, Iteral. *arXiv* **2012**, arXiv:1207.0152.
53. Abdel-Basset, M.; Abdel-Fatah, L.; Sangaiah, A.K. *Metaheuristic Algorithms: A Comprehensive Review*; Elsevier Inc.: London, UK, 2018; pp. 185–231. [CrossRef]
54. Opara, K.R.; Arabas, J. Differential Evolution: A survey of theoretical analyses. *Swarm Evol. Comput.* **2019**, *44*, 546–558. [CrossRef]
55. Mirjalili, S. Genetic algorithm. In *Evolutionary Algorithms and Neural Networks*; Springer: Cham, Switzerland, 2019; pp. 43–55.
56. Burke, E.K.; Hyde, M.; Kendall, G.; Ochoa, G.; Özcan, E.; Woodward, J.R. A Classification of Hyper-heuristic Approaches. In *Handbook of Metaheuristics*; Springer: Boston, MA, USA, 2010; pp. 449–468._15. [CrossRef]
57. Plata-González, L.F.; Amaya, I.; Ortiz-Bayliss, J.C.; Conant-Pablos, S.E.; Terashima-Marín, H.; Coello Coello, C.A. Evolutionary-based tailoring of synthetic instances for the Knapsack problem. *Soft Comput.* **2019**, *23*, 12711–12728. [CrossRef]
58. Epstein, S.L.; Petrovic, S. Learning a Mixture of Search Heuristics. In *Autonomous Search*; Hamadi, Y., Monfroy, É., Saubion, F., Eds.; Springer: Berlin/Heidelberg, Germany, 2012; pp. 97–127._5. [CrossRef]

Article

A Memetic Algorithm for the Cumulative Capacitated Vehicle Routing Problem Including Priority Indexes

Samuel Nucamendi-Guillén [†], Diego Flores-Díaz [†], Elias Olivares-Benitez [*,†] and Abraham Mendoza [†]

Facultad de Ingenieria, Universidad Panamericana, Álvaro del Portillo 49, Zapopan, Jalisco 45010, Mexico; snucamendi@up.edu.mx (S.N.-G.); 0198749@up.edu.mx (D.F.-D.); amendoza@up.edu.mx (A.M.)
* Correspondence: eolivaresb@up.edu.mx
† These authors contributed equally to this work.

Received: 27 April 2020; Accepted: 1 June 2020; Published: 5 June 2020

Abstract: This paper studies the Cumulative Capacitated Vehicle Routing Problem, including Priority Indexes, a variant of the classical Capacitated Vehicle Routing Problem, which serves the customers according to a certain level of preference. This problem can be effectively implemented in commercial and public environments where customer service is essential, for instance, in the delivery of humanitarian aid or in waste collection systems. For this problem, we aim to minimize two objectives simultaneously, the total latency and the total tardiness of the system. A Mixed Integer formulation is developed and solved using the AUGMECON2 approach to obtain true efficient Pareto fronts. However, as expected, the use of commercial software was able to solve only small instances, up to 15 customers. Therefore, two versions of a Memetic Algorithm with Random Keys (MA-RK) were developed to solve the problem. The computational results show that both algorithms provided good solutions, although the second version obtained denser and higher quality Pareto fronts. Later, both algorithms were used to solve larger instances (20–100 customers). The results were mixed in terms of quality but, in general, the MA-RK v2 consistently outperforms the first version. The models and algorithms proposed in this research provide useful insights for the decision-making process and can be applied to solve a wide variety of business situations where economic, customer service, environmental, and social concerns are involved.

Keywords: open vehicle routing; integer programming; split deliveries; logistic distribution; mixed integer formulations

MSC: [2010] 90C11; 90B06

---

1. Introduction

In this paper, we study the biobjective variant of the Cumulative Capacitated Vehicle Routing Problem (CCVRP), a "customer-centric" variant of the classical Capacitated Vehicle Routing Problem (CVRP) [1] in which a fleet of k vehicles serves a set of customers by respecting their priority, defined as an index assigned to each node to indicate the preference to be served. Unlike the traditional VRP, which focuses on the impact that routing costs have on logistics and, in particular, in the transportation activities within the supply chain, the CCVRP rises as a particularization that covers broad objectives centered around service level issues. This problem is relevant in contexts where both customer satisfaction and company profits are prioritized. Due to the importance of sustainable business practices nowadays, there is a need to develop distribution strategies aimed at reducing the negative impact that transportation activities have on the environment. An important application can also be found in the context of emergency logistics, where the distribution of medical aid becomes crucial, particularly when the distribution of different types of drugs or supplementary medical equipment have different levels of importance.

The CCVRP was introduced by Ngueveu et al. [2] as a generalization of the well known *k*-Traveling Repairmen Problem (*k*-TRP), with the objective of addressing problems in which customer metrics reflect the need for fast, equity and fair services. Since then, a diverse number of scenarios have been addressed: where a single vehicle can travel multiple trips [3,4], considering stochastic demand and split/unsplit deliveries [5], or when multiple depots are available [6]. Specifically, the CCVRP with priorities arises in the context of modeling situations in which customers have a differentiated level of attention and has attracted the interest of researchers over the past years due to its applicability in fields such as emergency logistics (i.e., level of damage), delivery logistics (i.e., delivery of different perishable items), personnel transportation, school bus transportation, machine scheduling considering due dates, maintenance operations including delivery dates, receiver wireless communications, etc.

As in the traditional VRP and its known variants, exact solution methods, heuristics, and metaheuristic algorithms are proposed to solve the CCVRP. Karagul et al. [7] pointed out that optimal solutions are possible to obtain in small-scale problems using exact solution methods. However, large-scale problems are difficult and time-consuming to solve to optimality. Hence, when it comes to real-life optimization, where problems are complex and deal with a significant amount of data, sometimes it is enough to find approximate solutions through the use of heuristic and metaheuristic methods.

Before the formal introduction of the problem in 2010, Kara et al. [8] developed a preliminary research studying a particular version of the CCVRP, named as CumVRP, in which the objective consists of minimizing the sumproduct of the arrival times and the demand of the node. Additionally, the CCVRP has been mainly studied from the monobjective function perspective. For this, several contributions involving mathematical models [9,10], exact algorithms [9,11], and heuristic and metaheuristic approaches [2,4,12–14] have been developed.

Previously, Sbihi and Eglese [15] dealt with the time-dependent routing problem (TDVRP) and applied this approach to green logistics. They established that the use of time-dependent models could obtain optimal solutions that produce less pollution by directing vehicles to roads where they can travel at faster speeds, which means away from congestion zones. However, this could represent traveling longer distances. Another benefit of this model is that it allows time window constraints to be satisfied more reliably. On the other hand, Kwon, Choi, and Lee [16] presented the heterogeneous vehicle routing problem that determines a set of vehicle routes that satisfies customer demands and vehicle capacities and minimize the sum of variable operating costs. An integer programming model was used, and for more complex problems, Tabu Search algorithms were developed to find solutions within a reasonable computational time. The results show that the amount of carbon emissions could be reduced by carbon trading without increasing the total costs.

Regarding the study of biobjective vehicle routing problems considering the customer-centric approach, few contributions can be found in the literature: the traveling repairman problem with profits (TRPP) [17], the *k*-Traveling Repairmen Problem with Profits (*k*-TRPP) under uncertainty [18], and a biobjective study of the minimum latency problem [19]. In the first work [17], the authors considered the *k*-TRP with the objective function that monotonically decreases over time. For this problem, they proposed a mathematical formulation. However, since the model should be solved separately for each path, the solution framework can be time-consuming, motivating the development of a Tabu-search procedure. The second research [18] addresses the *k*-TRP but enables the decision-maker to manage and control risk. For this, the authors proposed a mean-risk formulation and a heuristic approach based on an adaptive neighborhood search for the *k*-TRPP under uncertainty, providing high-quality results for small and medium-size instances. Finally, the most recent work [19] studies the Minimum Latency-Distance Problem (MLDP), analyzing the conflict between the total distance traveled by the vehicles and the total waiting time of the customers. The authors proposed a mathematical formulation and two heuristic procedures inspired on evolutionary algorithms, obtaining optimal results for instances up to 40 customers and extending the computational experimentation over the metaheuristics for instances up to 256 customers.

Considering the aforementioned discussion, to the best of our knowledge, no tailored approach has been proposed in the literature for the problem considering the risk-averse perspective. For this reason, in this paper ,we study the Biobjective Cumulative Capacitated Vehicle Routing Problem considering Profits (BCCVRP-Pr). To handle this new problem, we propose a mixed-integer formulation and a MA-RK procedure to efficiently deal with instances of reasonable size.

Elshaer and Awad [20] studied the diversity of variants of vehicle routing problems solved using metaheuristics, where eleven papers were identified using memetic algorithms. In a recent example, Li et al. [21] proposed a hybrid metaheuristic that combines a memetic algorithm, sequential variable neighborhood descent, and a revised 2-opt method, for the plug-in hybrid electric vehicle routing problem. In addition, Zhang et al. [22] developed a multiobjective memetic algorithm for the vehicle routing problem with time windows. To our knowledge, only Ngueveu et al. [2] applied memetic algorithms to the cumulative capacitated vehicle routing problem. They created two versions of their procedures to solve a single objective problem to minimize the sum of arrival times at the customers.

On the other side, memetic algorithms have been combined with random keys in some applications. One example is the hybridization of He et al. [23], where memetic algorithms were combined with a biased random key genetic algorithm and adaptive large neighbourhood search to solve a scheduling problem. Other applications to scheduling problems using memetic algorithms and random keys were proposed by Li and Yin [24] and Ghrayeb and Damodaran [25], among others. Although this combination of memetic algorithms and random keys have been used to solve sequencing problems, such as in the traveling salesman problem [26], this combination has not been used to solve complex routing problems such as the one presented in this paper.

The literature described above shows the novelty of the combination of memetic algorithms and random keys to solve complex multiobjective routing problems.

The remainder of this paper is organized as follows. Section 2 describes the proposed mathematical formulation, whereas Section 3 presents the algorithm developed. Section 4 reports the results obtained from the experimentation using a set of benchmark instances. Finally, Section 5 summarizes the major findings and proposes future research directions.

2. Mathematical Formulation

In this section, we present the formal definition of the BCCVRP-Pr as well as its corresponding mathematical formulation. First, the following parameters and variables are considered:

Parameters

- N: Number of customers
- K: Number of vehicles available

The BCCVRP-Pr can be formally defined as an undirected graph $G = (V, A)$ where $V = \{0, 1, 2, ..., N\}$ denote the node set and A is the set containing all arcs. The node 0 corresponds to the depot and the rest of the nodes form the set of customers $V' = \{1, 2, ..., N\}$. Each arc $(i, j) \in A$ has an associated travel time c_{ij} and each node $j \in V'$ has an associated demand d_j. A heterogeneous fleet of K vehicles is available, each with capacity $Q^k, k \in \{1, 2, ..., K\}$. It is assumed that the vehicle's overall capacity is enough to satisfy the total demand of all the clients. In addition, to consider the priority, a precedence matrix P is defined in which $p_{ij} = 1$ represents that customer i should be serviced before customer j and $p_{ij} = 0$ means that customer i can be served after the customer j.

The following additional parameters are considered:

- Q_{max}: Maximum capacity of any of the vehicles
- M: Maximum travel distance allowed (the same for all vehicles)

The goal is to find k tours that have only node 0 in the first position, and to cover all the nodes, while minimizing the sum of the latencies of the trips. The demand of all customers must be satisfied without exceeding the capacity of each vehicle. Customers should be served (preferably) according to their priority level, minimizing the total tardiness of the system.

For each path, the tardiness arises when a customer with lower priority is served before a customer with higher priority (even if both customers belong to different routes). In other words, the arrival time (t_i) for a customer with a lower priority index is less than the arrival time of a customer with a higher priority index (t_j). Qualitatively, tardiness is associated with customer dissatisfaction. However, because latency is estimated as a function of distance, then tardiness is obtained as the difference between their arrival times (when $p_{ij} = 1$), $l_{ij} = t_j - t_i$. Hence, the total tardiness of the system is computed, as shown in Equation (1):

$$Total\ tardiness = \sum_{i \in V'} \sum_{j \in V'} l_{ij} \tag{1}$$

Figures 1 and 2 graphically exhibit the conflict among the values of latency and tardiness for an instance that contains 12 nodes. The number above each node denotes the customer priority index (higher values indicate higher priorities). In Figure 1, the minimum total tardiness was obtained subject to the minimum total latency. Correspondingly, Figure 2 indicates the optimal total latency while assuring the minimum total tardiness.

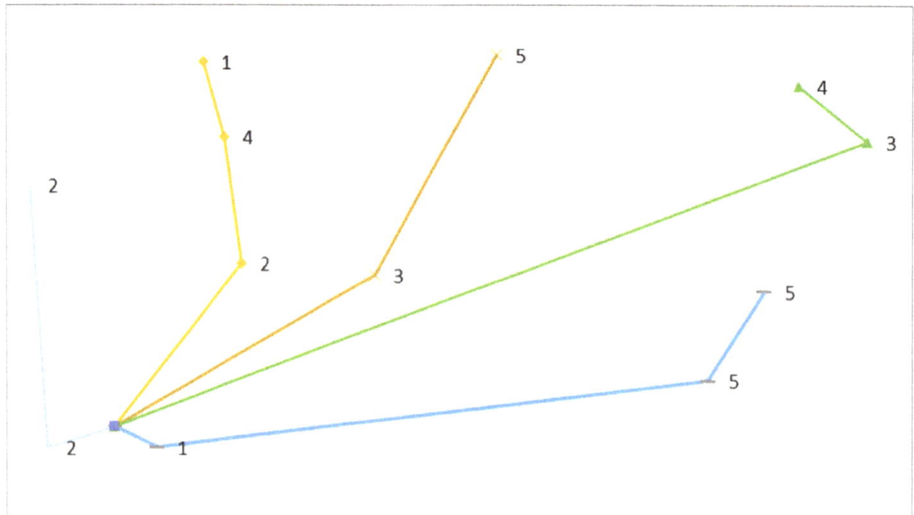

Figure 1. Solution routes minimizing the total latency: total latency = 1595.31; total tardiness = 4304.55.

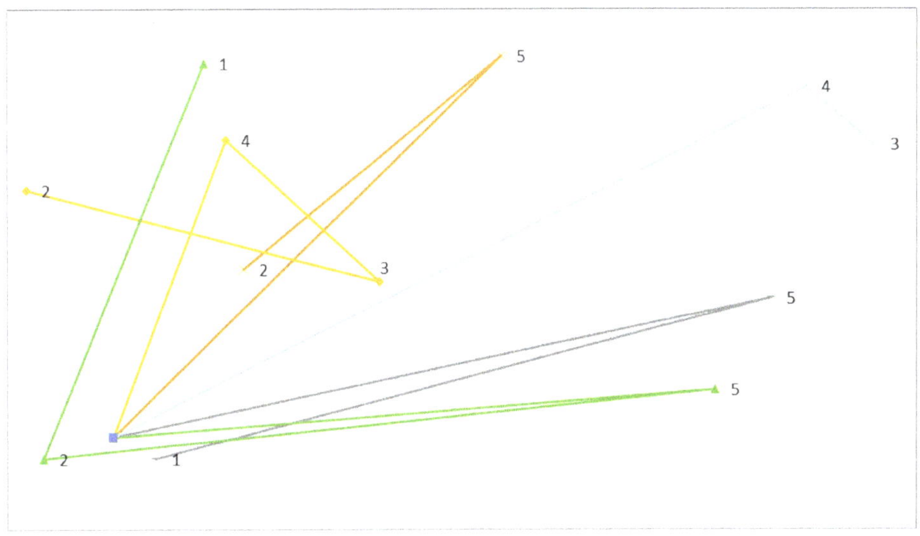

Figure 2. Solution routes minimizing the total tardiness: total latency = 2972.84; total tardiness = 0.

The formulation is based on the model presented in [10] for the classical CCVRP and following the single flow formulation proposed for the multiple traveling salesman problem [27].

Variables

$$w_{0j}^{k} = \begin{cases} 1, & \text{if the vehicle } k \text{ is assigned to a path from node } 0 \text{ to customer } j, \\ 0, & \text{otherwise.} \end{cases}$$

$$x_{ij} = \begin{cases} 1, & \text{if the arc } (i,j) \text{ is in the path of a vehicle,} \\ 0, & \text{otherwise.} \end{cases}$$

y_{ij} = number of remaining nodes after the node i on a route if $x_{ij} = 1$; 0, otherwise.

v_{0j}^{k} = the sum of demands of the nodes after node 0 on the route k if $w_{0j}^{k} = 1$; 0, otherwise.

v_{ij} = sum of demands of the nodes after node i on a route if $x_{ij} = 1$; 0, otherwise.

t_{0j}^{k} = Arrival time of node j from node 0 on a route k if $w_{0j}^{k} = 1$; 0, otherwise.

t_{ij} = Cumulative time at node j on a route if $x_{ij} = 1$; 0, otherwise.

I_{ij} = Tardiness in the arrival time to node i if node j is served first$(p_{ij} = 1)$; 0, otherwise.

The corresponding mathematical model is stated as follows:

$$min \ F_1 = L = \sum_{i \in V'} c_{0i} y_{0i} + \sum_{i \in V'} \sum_{\substack{j \in V' \\ j \neq i}} c_{ij} y_{ij}, \tag{2}$$

$$min \ F_2 = T = \sum_{i \in V'} \sum_{\substack{j \in V' \\ j \neq i}} I_{ij}, \tag{3}$$

subject to:

$$\sum_{j\in V'}\sum_{k\in K} w_{0j}^k = K, \tag{4}$$

$$\sum_{i\in V'} x_{i0} = K, \tag{5}$$

$$\sum_{j\in V'}\sum_{k\in K} w_{0j}^k = 1, \qquad \forall k \in K, \tag{6}$$

$$\sum_{k\in K} w_{0j}^k + \sum_{i\in V'} x_{ij} = 1, \qquad \forall j \in V', \tag{7}$$

$$\sum_{j\in V} x_{ij} = 1, \qquad \forall i \in V', \tag{8}$$

$$y_{0j} + \sum_{\substack{i\in V'\\i\neq j}} y_{ij} - \sum_{\substack{i\in V'\\i\neq j}} y_{ji} = 1, \qquad \forall j \in V', \tag{9}$$

$$y_{0j} \geq \sum_{k\in K} w_{0j}^k, \qquad \forall j \in V', \tag{10}$$

$$y_{ij} \geq x_{ij}, \qquad \forall i \in V'; \forall j \in V'; i \neq j, \tag{11}$$

$$y_{0j} \leq (N-K+1)\sum_{k\in K} w_{0j}^k, \qquad \forall j \in V', \tag{12}$$

$$y_{ij} \leq (N-K)\sum_{k\in K} x_{ij}, \qquad \forall i \in V'; \forall j \in V', \tag{13}$$

$$v_{0j}^k \geq d_j w_{0j}^k, \qquad \forall j \in V'; \forall k \in K, \tag{14}$$

$$v_{0j}^k \leq Q^k w_{0j}^K, \qquad \forall j \in V'; k \in K, \tag{15}$$

$$v_{ij} \geq d_j x_{ij}, \qquad \forall i \in V'; j \in V'; i \neq j, \tag{16}$$

$$v_{ij} \leq (Q_{max}-d_i)x_{ij}, \qquad \forall i \in V'; j \in V; i \neq j, \tag{17}$$

$$\sum_{k\in K} v_{0j}^k + \sum_{\substack{i\in V'\\i\neq j}} v_{ij} - \sum_{\substack{i\in V\\i\neq j}} v_{ji} = d_j, \qquad \forall j \in V', \tag{18}$$

$$t_{0j}^k = c_{0j} w_{0j}^k, \qquad \forall j \in V'; j \in V'; k \in K, \tag{19}$$

$$t_{0j}^k \leq M w_{0j}^k, \qquad \forall i \in V'; j \in V'; i \neq j; k \in K, \tag{20}$$

$$t_{ij} \geq c_{ij} x_{ij}, \qquad \forall i \in V'; j \in V'; i \neq j, \tag{21}$$

$$t_{ij} \leq (M-c_{ij})x_{ij}, \qquad \forall i \in V'; j \in V; i \neq j, \tag{22}$$

$$\sum_{\substack{h\in V\\h\neq i}} t_{ih} - \sum_{\substack{h\in V'\\h\neq i}} t_{hi} - \sum_{k\in K} t_{0i}^k = \sum_{\substack{j\in V\\j\neq i}} c_{ij} x_{ij}, \qquad \forall i \in V', \tag{23}$$

$$I_{ij} \geq p_{ij}\left(\sum_{k\in K} t_{0i}^k + \sum_{\substack{h\in V'\\h\neq i}} t_{hi} - \sum_{\substack{h\in V'\\h\neq j}} t_{hj} - \sum_{k\in K} t_{0j}^k\right), \qquad \forall i \in V'; j \in V'; i \neq j, \tag{24}$$

$$w_{0j}^{k} \in \{0,1\}, \qquad \forall j \in V'; \forall k \in K, \tag{25}$$

$$x_{ij} \in \{0,1\}, \qquad \forall i \in V'; \forall j \in V, \tag{26}$$

$$y_{ij} \geq 0, \qquad \forall i \in V; \forall j \in V, \tag{27}$$

$$v_{0j}^{k} \geq 0, \qquad \forall j \in V'; k \in K, \tag{28}$$

$$v_{ij} \geq 0, \qquad \forall i \in V'; j \in V, \tag{29}$$

$$t_{0j}^{k} \geq 0, \qquad \forall j \in V'; k \in K, \tag{30}$$

$$t_{ij} \geq 0, \qquad \forall i \in V'; j \in V, \tag{31}$$

$$I_{ij} \geq 0, \qquad \forall i \in V'; j \in V'. \tag{32}$$

In this formulation, the objective functions in Equations (2) and (3) seek a trade-off between the objective of the total latency and the total tardiness of the system. The constraints in Equations (4) and (5) ensure that exactly K arcs leave and return to the depot. The constraints in Equation (6) establish that a customer node can be visited for exactly one vehicle coming from the depot. Additionally, the constraints in Equations (7) and (8) impose that exactly one arc enters and leaves for each node associated with a customer. The constraints in Equations (10)–(13) help to avoid sub-tours and allow to calculate the position of the customer j on its respective route. The constraints in Equations (12) and (13) force variables y_{0j} and y_{ij} to be zero when $w_{0j}^{k} = 0$ and $x_{ij} = 0$, respectively. Regarding the capacity of the vehicles, the constraints in Equations (14) and (15) allow establishing lower and upper bounds for the cumulative demand of any route. These constraints are derived from a generalization of restrictions in Equations (10) and (11). The constraints in Equations (16) and (17) force variables v_{0j}^{k} and v_{ij} to be zero when variables $w_{0j}^{k} = 0$ and $x_{ij} = 0$, respectively. The constraints in Equation (18) ensure that the demand at each node i is fulfilled and in conjunction with Equations (16) and (17) estimate the load per vehicle. The constraints in Equations (19)–(23) control the arrival time to the nodes (customers). The constraints in Equation (24) estimate the tardiness (in case of violating the priority constraints). Finally, the constraints in Equations (25)–(32) establish the nature of the variables.

Reformulation Using Epsilon Constraint

In this subsection, we describe the characteristics of the model and the proposed reformulation. A fundamental task in multiobjective optimization is to find Pareto-optimal solutions. As a biobjective approach, we decided to implement a multiobjective method of resolution to generate an exact front of efficient solutions.

In the mathematical model, we note that the objective functions are separable. In other words, each of them involves different decision variables. On the one hand, y_{ij} allows estimating the total arrival time to the customers. On the other hand, I_{ij} computes the total tardiness for the case in which customers having a minor priority are served earlier than customers with higher priority index (regardless if they are located in the same route or belong to different routes).

To clarify this, the particular structure of the biobjective problem proposed herein is described below:

$$\min F_1 = L = \sum_{i \in V'} c_{0i} y_{0i} + \sum_{i \in V'} \sum_{\substack{j \in V' \\ j \neq i}} c_{ij} y_{ij} \tag{33}$$

$$\min F_2 = T = \sum_{i \in V'} \sum_{\substack{j \in V' \\ j \neq i}} I_{ij} \tag{34}$$

subject to :

$$Equations\,(4)–(24)$$

The aforementioned characteristics of the biobjective model are exploited by using an improved version of the ε-constraint method, named as AUGMECON2 [28], as a solution procedure. For every single routing decision in Equations (4)–(18), the minimum tardiness (min T) problem bounded by the constraints in Equations (19)–(24) is solved as principal objective function, transforming the latency function (L) into constraint. The results of the proposed method are shown in Section 4.

3. Metaheuristic Algorithm

This section is devoted to describe the metaheuristic approach capable of obtaining high quality solutions for small instances and able to deal with large size instances. The proposed method is based on a Memetic Algorithm (MA), a population procedure that has shown its effectiveness in solving sizeable combinatorial optimization problems by incorporating a local search procedure within a classical genetic algorithm. This procedure has been successfully applied for addressing the CCVRP [2], introducing efficient move evaluation procedures in operations $O(1)$ for some particular neighborhood structures. MAs have been also employed in solving other routing problems such as split delivery vehicle routing problems [29], capacitated location routing problems [30,31], vehicle routing problems with time windows [32], school bus routing problems [33], and green and healthcare routing problems [34,35].

3.1. Proposed Memetic Algorithm

Holland [36] was the first to propose Genetic Algorithms (GAs) inspired on ideas of evolution theory. Due to their simple and yet effective search procedure, several papers (e.g., [37]) describe their successful implementations in vehicle routing problems. In particular, Memetic algorithms (MAs) belong to the class of evolutionary algorithms that intensify the search by including local search within a classical genetic algorithm framework. According to Moscato and Cota [38], MAs are intrinsically concerned with exploiting all available knowledge about the problem under study. Due to this, a random key mechanism is included during the construction procedure in order to enhance the performance of the procedure. In this work, the MA proposed is adapted from the NSGA-II, successfully implemented in [39], and consists of the following procedures: initialization, recombination (crossover and local search), and classification in fronts. In our construction procedure, we include a random key that helps to generate a diverse initial population of feasible solutions of size N. Subsequently, during a predetermined number of successive generations (iterations), an offspring (P_t) of N individuals is generated from P_{t-1} through recombination and local search mechanisms, involving members representing tentative solutions (high-quality or non-dominated solutions) and members representing diverse solutions. After this, the individuals who belong to the previous and current generations are evaluated and grouped into fronts, according to the level of non-domination, as explained in [40]. To obtain the resulting offspring population of size N, the individuals are inserted into the set, starting with the one that belongs to the front of non-dominated solutions (F_0). Algorithm 1 shows the pseudocode of the overall MA with Random Keys procedure.

3.1.1. Constructive Procedure Based on Random Keys

The constructive procedure creates an initial population of feasible solutions based on generating a chain $S_a = \{1, 2, \cdots, n\}$ and, for each customer, an auxiliary random key R_a is used to encode the solution. Additionally, an empty set S_p is used to save the temporary assignment of the customers to the routes.

Algorithm 1: Memetic algorithm with random keys.

---

begin

    $it \leftarrow 0$;

    Initialize a population (P_0) of σ chromosomes implementing the constructive procedure based on random keys;

    Sort P_0 in fronts following the non-dominated sorting approach;

    for *(it = 1; it <= Maxiter; it + +)* **do**

        Generate an offspring population P_{it}, of size N, from P_{it-1}, using selection, crossover and local-search operators;

        Combine parent and offspring population $R_{it} = P_{it} \cap P_{it-1}$;

        Sort population using the non-dominated sorting approach, identify fronts $F_j = (1, 2, \cdots)$, and calculate the crowding distance for each solution in F_j;

        Make $T_{it+1} \leftarrow \varnothing, j \leftarrow 1$;

        while *($|T_{it+1}| + |F_j| < N$)* **do**

            $T_{it+1} \leftarrow T_{it+1} + F_j$;

            $j \leftarrow j + 1$;

            Sort solutions in F_j in decreasing order according to crowding distances, select the first $N - |T_{it+1}|$ elements of F_j and add it to T_{it+1};

        end

    end

end

Encoding mechanism: To encode the solution, a real number drawn randomly from $[0, 1)$ is assigned to every single position in R_a. Figure 3 depicts an example of this mechanism.

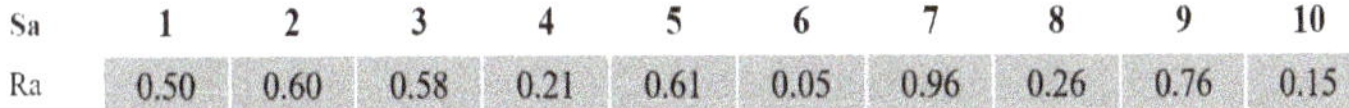

Figure 3. Example of the encoding mechanism.

Decoding mechanism: The decoding mechanism is applied based on the information of the random key R_a. The R_a chain is sorted in a non-decreasing order and their respective positions in S_a are sorted correspondingly. As a result, a random ordered chain S_a is obtained. Figure 4 exhibits the decoding mechanism.

Figure 4. Example of the decoding mechanism.

Assignment mechanism: In every iteration, the algorithm selects the corresponding *j*th customer in S_a and systematically tries to insert it into into a temporary set S_p in the first available position (procuring to maintain feasibility in the capacity of the vehicles). For instance, if in the first potential route, two customers have been previously assigned, the next open position will be the third one. In the case that the customer cannot be inserted in the route due to the capacity constraints, the insertion will be evaluated in the next available route. It is important to emphasize that, since the number of routes is given in advance, the construction procedure considers a parallel routing mechanism. In other words, it performs the evaluation of feasible insertions over all of the routes. If so, the algorithm continues by selecting the next customer at S_a. Otherwise, the construction mechanism stops. If the algorithm reaches a feasible assignment, then $S \leftarrow S_p$ and the solution is inserted into the population

Q_t. Otherwise, the algorithm destroys the partial constructed solution in S_p and generates a new random key R_a (to sort R_a).

The entire constructive procedure finishes when all of the customers have been assigned into S_p or after having a successive number of attempts without producing a feasible solution. When a feasible assignment is reached, the set S represents a feasible initial solution of routes. The customers already inserted in S are removed from S_a. Figure 5 shows an illustrative example of a feasible assignment.

Route 1	6	10	3	
Route 2	4	8	1	7
Route 3	2	5	9	

Figure 5. Example of a feasible assignment.

After reaching a feasible assignment, the sequencing mechanism is applied to construct each route by respecting the order of insertion and, based on this, the corresponding calculations of the objective functions L (representing the total latency of the system) and T (total tardiness of the system, based on the priorities of the customers) are performed. Algorithm 2 shows the pseudocode for this algorithm.

Algorithm 2: Constructive procedure (S, L, D).

```
begin
    Data: S ← ∅, Sp ← ∅, Sa =← ∅, L ← 0, T ← 0
    Fill chain in Sa = {1, 2, · · · , n};
    Create an auxiliary random key chain Ra with values [0, 1);
    Sort customers in Sa in a non-decreasing order according with their corresponding
       random values in Ra;
    j = 0;
    while (Sa ≠ ∅) do
        flag = 0;
        while Sa ≠ ∅ or flag = 1 do
            Select the jth customer from Sa;
            l = 0;
            if The insertion of the jth customer is feasible to insert in route l then
                Insert the jth customer in the lth route Sp;
                Remove the assigned customer from Sa;
                j++;
                flag = 0;
            end
            else if l < K then
                l++;
            end
            else
                Destroy the partial solution, Sp ← ∅;
                Establish Sa = {1, 2, · · · , n};
                Create a new random key Ra ;
                flag = 1 ;
            end
        end
        Compute the values of total latency L and total tardiness T for the individual;
    end
end
```

3.1.2. Crossover Procedure with Local Search Strategies

The proposed crossover procedure consists of the combination of two solutions, A and B, to create a new solution C. A tournament selection operator is incorporated to diversify the creation of new solutions. After obtaining the new individual C, a selective local search mechanism can be applied to improve it.

The procedure receives the following inputs: the current population (Q), the updated population (R), and the number of children to create (N). Then, the mechanism starts by selecting two individuals from the current generation (P_t). The first individual will belong to the front F_0, whereas the second one will be chosen at random from the entire population (generation). Subsequently, the customers of the routes for each solution are grouped into a single big chain by following the assignment order (starting from the first customer belonging to the first route and ending at the last customer of the latter one). As a result, the chains for the corresponding chromosomes A and B are obtained.

The creation of the new individual C is based on the information of the random keys (R_a) of each parent (chromosome). Then, a probability for inheriting is assigned to each parent. These probabilities are complementary. In other words, if the probability of $P_A = \alpha$ is assigned for the first chromosome (A), then the second chromosomes will receive a probability of $P_B = 1 - \alpha$. Then, for every position to fill in the R_a belonging to the customer, the roulette wheel is spun (RN) to determine if the element belonging to the R_a for the first or second parent must be selected to insert in the child. If the value of $RN \leq P_A$, the ith element R_a is included in the child. Otherwise, the element belonging to the second parent is selected. The mechanism stops when all of the positions have been evaluated. Figure 6 illustrates and example of this mechanism.

Parent A	1	2	3	4	5	6	7	8	9	10
P_A=0.6	0.50	0.60	0.58	0.21	0.61	0.05	0.96	0.26	0.76	0.15
RN	0.45	0.69	0.39	0.44	0.38	0.46	0.33	0.65	0.23	0.27
Child RK	0.50	0.43	0.58	0.21	0.61	0.05	0.96	0.60	0.76	0.15
P_b=0.6	0.72	0.43	0.73	0.77	0.08	0.68	0.25	0.60	0.62	0.69
Parent B	1	2	3	4	5	6	7	8	9	10

Figure 6. Example of child generation based on the roulette-tournament mechanism.

To enhance the creation of reasonable quality solutions, the probability assigned to the parent belonging to F_0 is always greater than 50%. Since the decoding procedure is a simple mechanism, it might occur that different random keys lead to an identical solution. Once the Ra for a child solution C is obtained, its feasibility is evaluated by calling back the constructive procedure. If the resulting solution is feasible, then the total latency L_C and tardiness T_C objectives are computed. On the contrary, the child is discarded, and a new second parent is selected (preserving the first individual) from P_t to restart the crossover procedure. Algorithm 3 depicts the pseudocode of the crossover mechanism.

Algorithm 3: Crossover procedure.

---

begin
> **Data:** Q, R
> **for** *(h = 1; h <= N; h + +)* **do**
> > Select randomly the first parent (A) from $F_0 \in Q$;
> > Select randomly the second parent (B) from Q;
> > Construct the big chain for each selected solution;
> > $i = 0$;
> > **repeat**
> > > Spin the wheel to obtain the value of *probability*;
> > > **if** *probability* $\leq \beta$ *and customer in A is available* **then**
> > > > Select the customer of the chromosome A;
> > >
> > > **end**
> > > **else if** *probability* $\leq \beta$ *and customer in B is available* **then**
> > > > Select the customer of chromosome B;
> > >
> > > **end**
> > > **else**
> > > > i++;
> > >
> > > **end**
> >
> > **until** *All positions in both parents have been evaluated*;
> >
> **end**
> **if** *feasible* **then**
> > Decode the corresponding solution for the new individual;
> > Compute the Latency (L) and Tardiness (T) values;
> > Spin the roulette to obtain a *rand* number;
> > **if** *rand* $\leq$ *threshold* **then**
> > > Apply the local search procedure over the individual C;
> >
> > **end**
> > Insert the new solution C in R.;
> >
> **end**
end

3.1.3. Local Search (LS) Procedure

The LS procedure is based on local search strategies, applied to intensify the search in pursuit of finding local minima. This procedure consists of five different neighborhood structures arranged into two classes, namely intra-route and inter-route mechanisms, performing them iteratively. This strategy has proved to be successful for a mono-objective version of the CCVRP [10]. Below, we describe each type of move:

- *Intra-route swap.* The procedure exchanges the positions of two customers belonging to the same route. For instance, if the customers to exchange belong to positions h and i, then arcs $(h-1, h)$, $(h, h+1)$, $(i-1, i)$ and $(i, i+1)$ are removed and replaced by arcs $(h-1, i)$, $(i, h+1)$, $(i-1, h)$ and $(h, i+1)$. It is important to remark that these movements do not affect feasibility in terms of capacity.
- *Intra-route reallocation.* This mechanism deletes a customer from its current position and reinserts it into another position on the same route.
- *Intra-route 2-opt.* In this operator, two non-adjacent edges $(h, h+1)$ and $(i, i+1)$ in the path $0, 1, 2, \ldots, h, h+1, \ldots, i, i+1, \ldots$ are deleted and replaced by (i, h) and $(h+1, i+1)$, resulting in the new path $0, 1, 2, \ldots, h, i, \ldots, h+1, i+1, \ldots$

- *Inter-routes interchange.* This strategy exchanges two customers belonging to different routes, as long as the move keeps feasibility (in terms of capacity).
- *Inter-routes reallocation.* For a given customer, the operator searches for the best position of the customer to move in any of the routes. If the best-identified position is different from the current one, the movement is performed.

The two major strategies operate as follows: At first, the initial solutions are sent to the intra-local search procedure, where the intra-local search strategies are applied. Then, the local minimum is forwarded to perform inter-routes local search strategies. These procedures are iteratively implemented, while the current solution value L keeps improving. In each process, the reallocation movement is performed first, and the execution of the interchange movement next. The first improvement criterion (*FI*) is used. Algorithm 4 exhibits this process.

Algorithm 4: Local search (S, L, T).

---

begin
  repeat
    $S^* = S$ and $L^* = L$, and $T^* = T$;
    S',L' T' $\leftarrow$ Intra-route local search (S,L, T);
    if *The solution is non-dominated* **then**
      $S = S'$ and $L = L'$, $T^* = T$;
    end
    S",L" T"$\leftarrow$ Inter-routes local search (S,L, T);
    if *The solution is non dominated* **then**
      $S = S''$ and $L = L''$, $T^* = T$;
    end
  until $L^* > L$;
  return S^*, L^*, T^*;
end

As observed, this mutation procedure seeks to insert improved individuals to the next generation, although the mechanism does not guarantee that the chromosome selected can be deeply improved. Because the size of this subset is relatively small, it is always possible to find a chromosome to be improved.

This procedure differentiates the two versions of the MA. For the first version (MA-RK v1), all of the feasible individuals generated by the crossover mechanism are sent to the LS procedure (*threshold* = 1). For the second version (MA-RK v2), the local search mechanism is applied only to a certain percentage of individuals, expecting to accelerate the performance of the algorithm in terms of CPU time.

4. Computational Results

This section is devoted to reporting the computational experiments conducted to assess the efficiency of the proposed approach. First, we provide the set of instances used to perform the tests, as well as the characteristics of the computational equipment used. Secondly, we present the parameter setting for our versions of the MA. Finally, the experimental results for both the mathematical formulation and the metaheuristic procedure are displayed, accompanied by the respective discussion.

4.1. Test Instances

The instances used to conduct the experimentation were adapted from the ones used in the literature to evaluate the multi depot VRP with heterogeneous fleet: Koulaeian et al. [41] (Kou15), Chunyu and Xiaobo [42] (CaX10), Gillett and Jhonson [43] (GaJ76-7–GaJ76-12), and Augerat et al. [44] (Pn16k8 and Pn23k8). Even though there are some instances proposed by Talliard [45] and Li et al. [46] for the classical Heterogeneous Fleet Vehicle Routing Problem, we decided to use these instances since some of them provide a reasonable size in the number of customers for assessing the model.

The size of the instances ranges from 12 to 100 nodes and from 2 to 10 vehicles. The generated problem instances are characterized by the following criteria: (i) number of customers; (ii) number of vehicles; (iii) coordinates (x,y) for all locations (including the depot); (iv) demand of each node; and (v) priority index for each node. Since the original instances consider multiple depots and do not consider any preference index between the customers to be served, we selected the first depot as the single-origin, and we included a priority parameter by assigning a numerical index within $\{1, \lceil \sqrt{n} \rceil\}$, based on a uniform distribution. The customers having the highest value of the index represent the ones that should be first served (highest level of priority).

Since the modified instances are set to deal with a different problem, and to facilitate the report of the results, we decided to rename them using the nomenclature "FNO-x", where x denotes a consecutive number assigned according to the rank assigned to the instance (in terms of the number of nodes, following a non-increasing order). For example, the instance Kou15 is the one with the lowest number of customers (12); therefore, it was renamed as FNO1. The remaining instances based on Pn16k8, CaX10, Pn23k8, GaJ76-7, GaJ76-8, GaJ76-9, GaJ76-11, and GaJ76-12 were renamed as FNO2, FNO3, FNO4, FNO5, FNO6, FNO7, FNO8, and FNO9 respectively. In the case of the instance GaJ76-10, it was named FNO10 because it has the largest size in the number of customers. These new instances are available by request.

All of the experiments were conducted using a PC Intel Core i7 @2.30 GHz with 16 GB of RAM Memory under Windows 10. The formulation was modeled using AMPL and solved using Gurobi 9.0. For each instance, we established a time limit of 7200 s (2 h). In the case of the MA-RK, both versions were coded using the C++ language. In the next subsection, the parameter tuning is presented.

4.2. Parameters Setting

In the case of the MA-RK v1, the values for the parameters corresponding to the size of the population N, the threshold value β, and the maximum number of generations D were set as 1000, 0.1, and 100, respectively. In the case of MA-RK v2, we set a threshold of 0.4. These values were obtained after performing a preliminary analysis over a subset of instances randomly selected. In addition, several tests with different number of iterations were conducted, finding that, for all the analyzed instances, the MA-RK stops improving when it reaches 80% of the maximum number of iterations, depending on the instance. Lastly, to evaluate consistency, each instance was executed 10 times, and the best front obtained is reported. The next sections show the numerical results computed over the test instances.

Experimental Results

The first set of experiments aims at evaluating the performance of the formulation concerning optimality (optimally solved instances), and the effectiveness of the MA-RK comparing the results with those obtained by the resolution of the model. We first present the results for Instances FNO1 and FNO2 (up to 15 nodes).

The following metrics were used to compare the performance of the exact and approximation procedures:

- Number of points on the front (the larger, the better)
- CPU time (in seconds) (the shorter, the better)
- k-distance [47] (the smaller, the better)

- Hypervolume [48] (the larger, the better)
- The coverage of the fronts [48] computed of one front over another, denoted by c(X', X") (the higher, the better)

These metrics have shown their successful implementation in biobjective VRPs [19,39].

The number of non-dominated points measures the ability of each method to find efficient fronts. Table 1 summarizes these results for Instances FNO1 and FNO2. Figures 7 and 8 show the Pareto front for each instance and method.

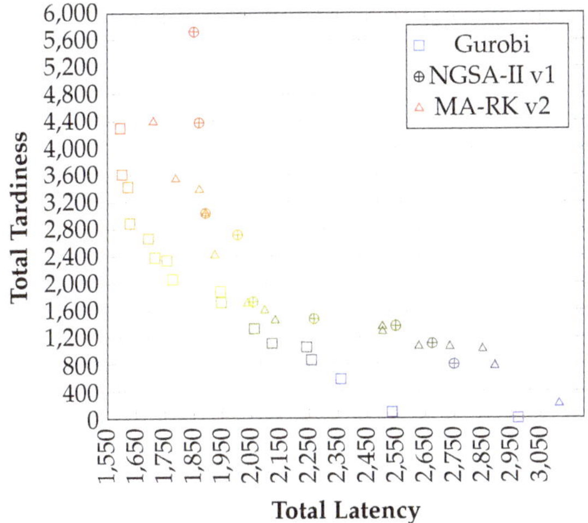

Figure 7. Pareto front for Instance FNO1.

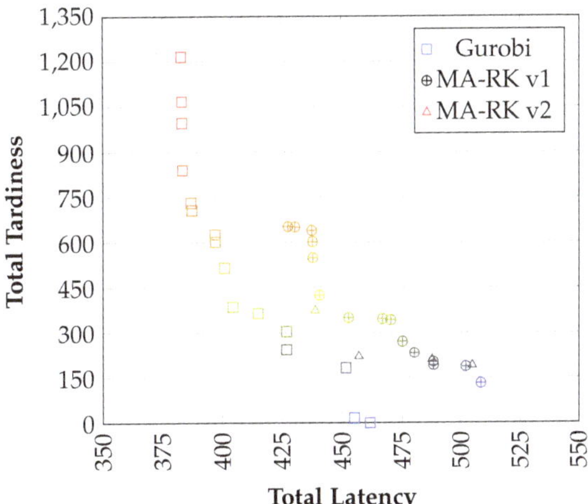

Figure 8. Pareto front for Instance FNO2.

A point to highlight in Figures 7 and 8 and Table 1 is that, for both instances, the Pareto front obtained by Gurobi is densely crowded. Additionally, notice that both versions of MA-RK performed differently over the solved instances. In particular, for Instance FNO1, the second version of the algorithm produced a front that is closer to the optimal front obtained by CPLEX. On the contrary, for Instance FNO2, both algorithms produced fronts near to the optimal front and, in particular, the MA-RK v1 produced a more dense front than the MA-RK v2.

Table 1. Quantity of non-dominated points for Instances FNO1 and FNO2.

Instance Name	n	k	Gurobi	MA-RK v1	MA-RK v2
FNO1	12	5	17	9	15
FNO2	15	8	16	15	4

Regarding the computational time, the summarized results are presented in Table 2. In this case, the exact method required around 1 h for solving the FNO1 instance (12 nodes), whereas the time required to solve the FNO2 (15 nodes) instance was almost four times as long. In particular, both versions of the metaheuristic required less than 1 s to obtain the solutions. This fact, in conjunction with the metric of the quantity of non-dominated points, supports the evidence that, in general, the MA-RK algorithm performed very well.

Table 2. Elapsed CPU time (in seconds) for Instances FNO1 and FNO2.

Instance Name	Gurobi	MA-RK v1	MA-RK v2
FNO1	3,641.179	0.177	0.125
FNO2	13,742.185	0.256	0.149

Regarding the density of the fronts, Table 3 shows the average k-distance value of all points on the efficient frontiers for each instance, with $k = 3$. Specifically, the MA-RK produced fronts with more density than AUGMECON2. In particular, for the FNO2 instance, it can be seen that the NSGA-v1 obtained the minimum values for the maximum and average distances, while the MA-RK v2 obtained a denser front for Instance FNO1.

Table 3. Maximum and average k-distances for the FNO1 and FNO2.

Instance Name	Exact		MA-RK v1		MA-RK v2	
	Max	Avg	Max	Avg	Max	Avg
FNO1	0.41624	0.117976	0.469738	0.211416	0.225514	0.106864
FNO2	0.180034	0.121934	0.167452	0.0844987	0.410824	0.320028

To verify the efficiency of the MA-RK (in both versions), we used the hypervolume metric. Table 4 displays the obtained results. Again, the MA-RK v2 provided a better value of hypervolume for Instance FNO1, while the MA-RK v1 performed better on Instance FNO2.

Table 4. Hypervolume for Instances FNO1 and FNO2.

Instance Name	Exact	MA-RK v1	MA-RK v2
FNO1	0.821569	0.596382	0.64612
FNO2	0.799919	0.454889	0.432928

Finally, we used the coverage measure (considering only the strict domination). Tables 5 and 6 exhibit the results. In these tables, a value of $C(X', X'')$ equal to 1 means that all points in the estimated efficient frontier X'' are strictly dominated by points in the estimated efficient frontier X'. As expected, the exact method dominates both algorithms entirely in terms of the space covered. Regarding the metaheuristic procedures, for Instance FNO1, the MA-RK v2 was able to generate points that dominate almost 77.78% over the ones generated by the MA-RK v1. For Instance FNO2, the front of the MA-RK v2 dominates 33.33% of the points generated by the MA-RK v1 which, in turn, dominates 25% of the points generated by the MA-RK v2.

Table 5. Coverage of two sets value for Instance FNO1.

X'/X''	Exact	MA-RK v1	MA-RK v2
Exact	0	1	1
MA-RK v1	0	0	0.066
MA-RK v2	0	0.778	0

Table 6. Coverage of two sets value for Instance FNO2.

X'/X''	Exact	MA-RK v1	MA-RK v2
Exact	0	1	1
MA-RK v1	0	0	0.25
MA-RK v2	0	0.333	0

We observe that, for Instances FNO1 and FNO2, when the minimum value of latency is obtained, the maximum amount of total tardiness rises to 3.2 times the cost of latency, which might translate to a higher level of customer dissatisfaction. On the contrary, when the minimum value of total tardiness is reached, the overall latency of the system rises up to 1.4 times the optimal (minimum) value. In this case, the increment of latency translates into a significant increase in the cost and, therefore, in a reduction of profit. In addition, it is obvious that the prioritization of the customers generates an unbalance in their demand, especially for the case of having routes where relatively few customers can have significantly high amounts of demand compared to the rest.

Another aspect to highlight is that the decision-making process can be seen from two perspectives: (1) savings in tardiness costs can represent up to 72% of the total costs; and (2) savings in latency produce savings for up to 28% of the total costs. In other words, according to the objective function, if a preference must be defined a priori, tardiness must be more important than latency.

4.3. Experimental Results for Larger Instances

The experimentation considering large size instances was conducted in both versions of the algorithm. The complementary experimentation involves instances of up to 100 nodes. Tables 7–12 display the results of our computational experimentation.

In Table 7, Column 1 displays the name of the instance. Columns 2 and 3 indicate the size in terms of the number of nodes and the number of routes. Columns 4 and 5 report the number of non-dominated solutions obtained by each algorithm. For Tables 9 and 10, Column 1 shows the name of the instance, whereas Columns 2 and 3 report the value of the corresponding algorithms over the evaluated metric. Specifically, for Table 11, two columns are used to indicate the maximum and average k-distances for each procedure.

Following the same sequence used in the previous section, the first metric to compare is the quantity of nondominated points. Table 7 reports the number of non-dominated points obtained by each algorithm (Pareto front). According to the information there, MA-RK v2 was able to obtain a higher quantity of non-dominated points. In some instances, the number of points reported almost doubled the amount of the ones obtained by MA-RK v1.This can be explained by the fact that MA-RK v2 generates more diverse solutions, since the process of intensification is selective.

Table 7. Quantity of non-dominated points for large-size instances.

Instance Name	n	k	MA-RK v1	MA-RK v2
FNO3	20	3	8	5
FNO4	22	8	8	6
FNO5	75	4	9	11
FNO6	75	7	5	18
FNO7	75	10	7	13
FNO8	75	7	10	5
FNO9	75	8	13	11
FNO10	100	5	8	12

In Table 8, the minimum and maximum values for each objective are shown. According to the information obtained, for most of the instances, the MA-RK v2 reports better values for the total latency. In addition, the MA-RK v2 produces better values of tardiness for most of the instances. In summary, the selective version of the MA-RK clearly outperforms the MA-RK v1.

Table 8. Minimum and maximum values for both objectives functions for large-size instances.

Instance Name	Type of Objective	MA-RK v1		MA-RK v2	
		Min	Max	Min	Max
FNO3	Latency	5327.35	7987.42	**4852.72**	**7335.90**
	Tardiness	4853.59	19,782.90	**3296.29**	**11,060.70**
FNO4	Latency	**660.85**	1102.98	759.182	**975.53**
	Tardiness	**233.65**	**1159.62**	290.55	2072.82
FNO5	Latency	8275.60	**10,376.60**	6788.84	10,460.50
	Tardiness	27,515.3	66,396.30	**23,903.30**	**51,934.70**
FNO6	Latency	7366.98	**9251.54**	6933.74	16,841.5
	Tardiness	20,883.80	31,319.70	**11,558.3**	**16,063.70**
FNO7	Latency	12,511.70	15,026.20	**10,027.10**	**11,253.50**
	Tardiness	82,739.9	**124,013.00**	67,365.4	144,892
FNO8	Latency	13,683.50	16,203.90	**12,833.5**	**14,208.1**
	Tardiness	116,976.00	199,977.00	**101,302**	**154,659**
FNO9	Latency	9617.94	14,949.70	**9150.29**	**11,955.2**
	Tardiness	69,126.70	130,754.00	**58,969.5**	**123,515.00**
FNO10	Latency	31,092.50	40,164.90	**26,484.2**	**35,107.6**
	Tardiness	343,722.00	530,071.00	**272,748.00**	**335,684.00**

Regarding the performance of the algorithms, it can be noticed that, for larger instances, the MA-RK v2 clearly outperforms the MA-RK v1. To better illustrate this, the fronts of Instances FNO8 and FNO10 are displayed in Figures 9 and 10.

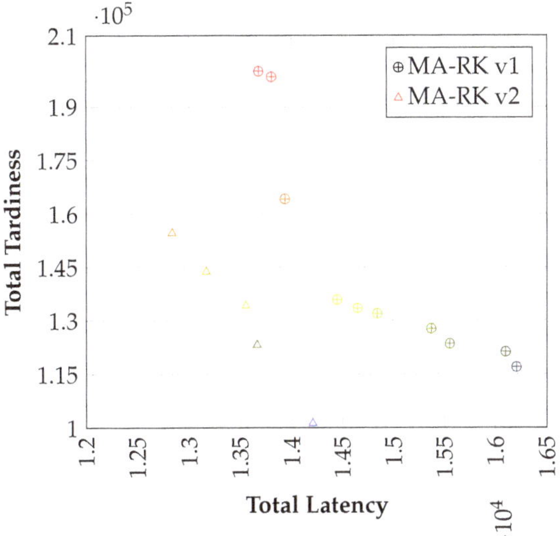

Figure 9. Pareto fronts for Instance FNO8.

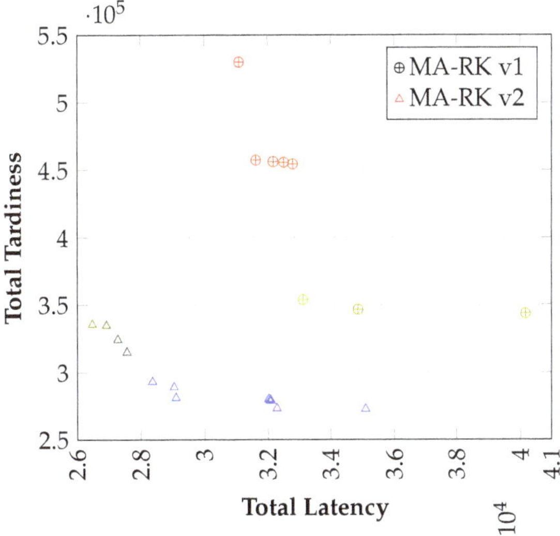

Figure 10. Pareto fronts for Instance FNO10.

Due to this, the metric of the execution time was evaluated to verify if any of the versions performs more quickly. Table 9 displays the elapsed CPU time for the best execution.

Table 9. Elapsed CPU time for the rest of the instances.

Instance Name	MA-RK v1	MA-RK v2
FNO3	0.262	0.817
FNO4	0.768	1.087
FNO5	6.415	6.586
FNO6	9.919	11.837
FNO7	62.262	70.162
FNO8	47.039	47.819
FNO9	48.816	54.017
FNO10	96.541	97.29

As expected, the required time increases as the size of the instances increased. However, both versions of the metaheuristic were working within a similar computational performance range.

As for the third metric, the hypervolume, the results of the algorithm are displayed in Table 10. There, we do not have enough evidence to confirm that any algorithm outperforms the other. What can be confirmed is that MA-RK v2 produced higher values of hypervolume for seven out of eight instances. However, for the instance where the MA-RK v1 obtained better values, the difference against the NSGA was small.

Table 10. Hypervolume for the rest of the instances.

Instance Name	MA-RK v1	MA-RK v2
FNO3	0.680387	0.836718
FNO4	0.752206	0.648306
FNO5	0.397265	0.823822
FNO6	0.642281	0.929850
FNO7	0.472473	0.646997
FNO8	0.433841	0.847476
FNO9	0.527335	0.836342
FNO10	0.390825	0.959196

Regarding the density of the frontiers, Table 11 shows the results obtained. From these results, it can be noticed that, in most of the cases, the MA-RK v2 produced lower values for the maximum distances (more compactness). However, the MA-RK v1 algorithm produced better values for the average distances.

Table 11. Maximum and average k-distances for the large-size instances.

Instance Name	MA-RK v1		MA-RK v2	
	Max	Avg	Max	Avg
FNO3	0.579687	0.308544	0.503147	0.389821
FNO4	0.418041	0.256448	0.643338	0.344496
FNO5	0.483586	0.166288	0.742506	0.168127
FNO6	0.516558	0.298917	0.483949	0.107271
FNO7	0.342773	0.179116	0.628917	0.176940
FNO8	0.371050	0.212086	0.386249	0.248088
FNO9	0.606253	0.161301	0.574856	0.179261
FNO10	0.387176	0.187895	0.221685	0.0603717

Finally, Table 12 reports the values obtained for the set coverage metric. The first column refers to the name of the instance, and the rest show comparisons in coverage between the algorithms. Again, MA-RK v2 performed better than MA-RK v1, by dominating the entire front provided by MA-RK v1. This confirms that the selective version of the MA-RK clearly dominates.

Table 12. Coverage of two sets value for the rest of the instances.

Instance Name	X'/X"	Exact	
		MA-RK v1	**MA-RK v2**
FNO3	MA-RK v1	0	0
	MA-RK v2	0.875	0
FNO4	MA-RK v1	0	0.333
	MA-RK v2	0.500	0
FNO5	MA-RK v1	0	0
	MA-RK v2	0.889	0
FNO6	MA-RK v1	0	0
	MA-RK v2	0.800	0
FNO7	MA-RK v1	0	0
	MA-RK v2	1	0
FNO8	MA-RK v1	0	0
	MA-RK v2	1	0
FNO9	MA-RK v1	0	0
	MA-RK v2	1	0
FNO10	MA-RK v1	0	0
	MA-RK v2	1	0

In summary, we conclude that, even when both metaheuristics provide good results in a reasonable computational time, the MA-RK v2 consistently outperforms the non-selective MA-RK version.

5. Conclusions and Future Work

This study addressed the biobjective Cumulative Capacitated Vehicle Routing Problem. This problem mainly arises in commercial contexts such as the delivery of perishable goods, in which there are differentiated based on priority indexes. In the case of a pooled transportation service, it might help to estimate the trade-off between delivering the orders in the same sequence as customers board the vehicle and the minimum arrival time of the system. For this problem, a mixed-integer programming formulation and two metaheuristic algorithms were developed. A commercial optimization software was able to solve the model for small size instances, whereas the algorithms showed their effectiveness by providing feasible results in a reasonable amount of computational time.

The algorithms showed their efficiency by providing good quality fronts for the small size instances. Additionally, for larger instances, both algorithms provided good values for the multiobjective metrics evaluated. Although none of the algorithms outperformed each other, the MA-RK v2 obtained fronts with a higher quantity of points, more density, and more coverage of the sets. However, the MA-RK v1 was slightly faster. One intuition is that MA-RK v1 is more intensive in the local search, stagnating in local optima, while MA-RK v2 maintains the diversity, allowing to escape from local optima and populating the Pareto-fronts. However, more computational experiments are needed to clarify this effect.

In summary, all procedures provided a positive contribution to a more sustainable balance between economic and customer service objectives. Our results provide useful insights for business applications in terms of considering customer satisfaction and gaining a valuable sustainable advantage given that the reduction in the traveled time translates into a reduction of CO_2 emissions.

Future research lines include the design of routes using congested environments with travel speed variation. This fact can be addressed either by modifying the objective function to include the variability in the travel time during the day or using a risk aversion approach, by adding a profit to each node associated with the order of visit. In addition, considerations involving time windows as priority metrics, and demand uncertainty can be worth consideration, as well as factors such as labor costs or balance of the total traveled distance among routes, which seem to dominate the overall cost.

Author Contributions: Conceptualization, S.N.-G. and E.O.-B.; methodology, S.N.-G. and D.F.-D.; software, S.N.-G. and D.F.-D.; validation, S.N.-G., D.F.-D., and E.O.-B.; formal analysis, S.N.-G., D.F.-D., and E.O.-B.; investigation, S.N.-G. and D.F.-D.; resources, S.N.-G. and E.O.-B.; data curation, S.N.-G. and D.F.-D.; writing—original draft preparation, S.N.-G.; writing—review and editing, E.O.-B. and A.M.; visualization, S.N.-G.; supervision, S.N.-G. and E.O.-B.; project administration, S.N.-G. and E.O.-B.; and funding acquisition, S.N.-G., E.O.-B., and A.M. All authors have read and agreed to the published version of the manuscript.

Funding: This work was supported by Universidad Panamericana through the grant "Fomento a la Investigación 2019", under project code UP-CI-2019-ING-GDL-08.

Acknowledgments: This work was supported by the Universidad Panamericana through the grant "Fondo Fomento a la Investigación UP 2019", under project code UP-CI-2019-ING-GDL-08.

Conflicts of Interest: The authors declare no conflict of interest.

References

1. Dantzig, G.B.; Ramser, J.H. The truck dispatching problem. *Manag. Sci.* **1959**, *6*, 80–91. [CrossRef]
2. Ngueveu, S.U.; Prins, C.; Calvo, R.W. An effective memetic algorithm for the cumulative capacitated vehicle routing problem. *Comput. Operations. Res.* **2010**, *37*, 1877–1885. [CrossRef]
3. Martínez-Salazar, I.; Angel-Bello, F.; Alvarez, A. A customer-centric routing problem with multiple trips of a single vehicle. *J. Oper. Res. Soc.* **2015**, *66*, 1312–1323. [CrossRef]
4. Rivera, J.C.; Afsar, H.M.; Prins, C. A multistart iterated local search for the multitrip cumulative capacitated vehicle routing problem. *Comput. Optim. Appl.* **2015**, *61*, 159–187. [CrossRef]
5. Gaur, D.R.; Mudgal, A.; Singh, R.R. Improved approximation algorithms for cumulative VRP with stochastic demands. *Discret. Appl. Math.* **2018**. [CrossRef]
6. Lalla-Ruiz, E.; Voß, S. A POPMUSIC approach for the Multi-Depot Cumulative Capacitated Vehicle Routing Problem. *Optim. Lett.* **2020**, *14*, 671–691. [CrossRef]
7. Karagul, K.; Sahin, Y.; Aydemir, E.; Oral, A. A Simulated Annealing Algorithm Based Solution Method for a Green Vehicle Routing Problem with Fuel Consumption. In *Lean and Green Supply Chain Management: Optimization Models and Algorithms*; Paksoy, T., Weber, G.W., Huber, S., Eds.; Springer International Publishing: Cham, Switzerland, 2019; pp. 161–187, doi:10.1007/978-3-319-97511-5_6. [CrossRef]
8. Kara, İ.; Kara, B.Y.; Yetiş, M.K. Cumulative vehicle routing problems. In *Vehicle Routing Problem*; IntechOpen: Rijeka, Croatia, 2008.
9. Rivera, J.C.; Afsar, H.M.; Prins, C. Mathematical formulations and exact algorithm for the multitrip cumulative capacitated single-vehicle routing problem. *Eur.J. Oper. Res.* **2016**, *249*, 93–104. [CrossRef]
10. Nucamendi-Guillén, S.; Angel-Bello, F.; Martínez-Salazar, I.; Cordero-Franco, A.E. The cumulative capacitated vehicle routing problem: New formulations and iterated greedy algorithms. *Expert Syst. Appl.* **2018**, *113*, 315–327. [CrossRef]
11. Lysgaard, J.; Wøhlk, S. A branch-and-cut-and-price algorithm for the cumulative capacitated vehicle routing problem. *Eur. J. Oper. Res.* **2014**, *236*, 800–810. [CrossRef]
12. Ribeiro, G.M.; Laporte, G. An adaptive large neighborhood search heuristic for the cumulative capacitated vehicle routing problem. *Comput. Oper. Res.* **2012**, *39*, 728–735. [CrossRef]
13. Ozsoydan, F.B.; Sipahioglu, A. Heuristic solution approaches for the cumulative capacitated vehicle routing problem. *Optimization* **2013**, *62*, 1321–1340. [CrossRef]
14. Ke, L.; Feng, Z. A two-phase metaheuristic for the cumulative capacitated vehicle routing problem. *Comput. Oper. Res.* **2013**, *40*, 633–638. [CrossRef]
15. Sbihi, A.; Eglese, R.W. Combinatorial optimization and Green Logistics. *4OR* **2007**, *5*, 99–116, doi:10.1007/s10288-007-0047-3. [CrossRef]

16. Kwon, Y.J.; Choi, Y.J.; Lee, D.H. Heterogeneous fixed fleet vehicle routing considering carbon emission. *Transp. Res. Part D Transp. Environ.* **2013**, *23*, 81–89, doi:10.1016/j.trd.2013.04.001. [CrossRef]

17. Dewilde, T.; Cattrysse, D.; Coene, S.; Spieksma, F.C.; Vansteenwegen, P. Heuristics for the traveling repairman problem with profits. *Comput. Oper. Res.* **2013**, *40*, 1700–1707. [CrossRef]

18. Bruni, M.; Beraldi, P.; Khodaparasti, S. A heuristic approach for the k-traveling repairman problem with profits under uncertainty. *Electron. Notes Discret. Math.* **2018**, *69*, 221–228. [CrossRef]

19. Arellano-Arriaga, N.A.; Molina, J.; Schaeffer, S.E.; Álvarez-Socarrás, A.; Martínez-Salazar, I.A. A biobjective study of the minimum latency problem. *J. Heuristics* **2019**, *25*, 431–454. [CrossRef]

20. Elshaer, R.; Awad, H. A taxonomic review of metaheuristic algorithms for solving the vehicle routing problem and its variants. *Comput. Ind. Eng.* **2020**, *140*, doi:10.1016/j.cie.2019.106242. [CrossRef]

21. Li, X.; Shi, X.; Zhao, Y.; Liang, H.; Dong, Y. SVND enhanced metaheuristic for plug-in hybrid electric vehicle routing problem. *Appl. Sci.* **2020**, *10*, 441, doi:10.3390/app10020441. [CrossRef]

22. Zhang, K.; Cai, Y.; Fu, S.; Zhang, H. Multiobjective memetic algorithm based on adaptive local search chains for vehicle routing problem with time windows. *Evol. Intell.* **2019**, doi:10.1007/s12065-019-00224-7. [CrossRef]

23. He, L.; Guijt, A.; de Weerdt, M.; Xing, L.; Yorke-Smith, N. Order acceptance and scheduling with sequence-dependent setup times: A new memetic algorithm and benchmark of the state of the art. *Comput. Ind. Eng.* **2019**, *138*, doi:10.1016/j.cie.2019.106102. [CrossRef]

24. Li, X.; Yin, M. A hybrid cuckoo search via Lévy flights for the permutation flow shop scheduling problem. *Int. J. Prod. Res.* **2013**, *51*, 4732–4754, doi:10.1080/00207543.2013.767988. [CrossRef]

25. Ghrayeb, O.; Damodaran, P. A hybrid random-key genetic algorithm to minimize weighted number of late deliveries for a single machine. *Int. J. Adv. Manuf. Technol.* **2013**, *66*, 15–25, doi:10.1007/s00170-012-4302-1. [CrossRef]

26. Samanlioglu, F.; Ferrell, W.; Kurz, M. An interactive memetic algorithm for production and manufacturing problems modelled as a multiobjective travelling salesman problem. *Int. J. Prod. Res.* **2012**, *50*, 5671–5682, doi:10.1080/00207543.2011.593578. [CrossRef]

27. Gavish, B.; Graves, S.C. *The Travelling Salesman Problem and Related Problems*; Massachusetts Institute of Technology, Operations Research Center: Cambridge, MA, USA, 1978.

28. Mavrotas, G.; Florios, K. An improved version of the augmented ε-constraint method (AUGMECON2) for finding the exact pareto set in multiobjective integer programming problems. *Appl. Math. Comput.* **2013**, *219*, 9652–9669, doi:10.1016/j.amc.2013.03.002. [CrossRef]

29. Boudia, M.; Prins, C.; Reghioui, M. An effective memetic algorithm with population management for the split delivery vehicle routing problem. In *International Workshop on Hybrid Metaheuristics*; Springer: Berlin/Heidelberg, Germany, 2007; pp. 16–30.

30. Karaoglan, I.; Altiparmak, F. A memetic algorithm for the capacitated location-routing problem with mixed backhauls. *Comput. Oper. Res.* **2015**, *55*, 200–216. [CrossRef]

31. Kechmane, L.; Nsiri, B.; Baalal, A. A memetic algorithm for the capacitated location-routing problem. *Int. J. Adv. Comput. Sci. Appl.* **2016**, *7*. [CrossRef]

32. Nalepa, J.; Blocho, M. Adaptive memetic algorithm for minimizing distance in the vehicle routing problem with time windows. *Soft Comput.* **2016**, *20*, 2309–2327. [CrossRef]

33. Sales, L.P.A.; Melo, C.S.; Bonates, T.O.; Prata, B.A. Memetic algorithm for the heterogeneous fleet school bus routing problem. *J. Urban Plan. Dev.* **2018**, *144*, 04018018. [CrossRef]

34. Decerle, J.; Grunder, O.; El Hassani, A.H.; Barakat, O. A memetic algorithm for a home health care routing and scheduling problem. *Oper. Res. Health Care* **2018**, *16*, 59–71. [CrossRef]

35. Peng, B.; Zhang, Y.; Gajpal, Y.; Chen, X. A Memetic Algorithm for the Green Vehicle Routing Problem. *Sustainability* **2019**, *11*, 6055. [CrossRef]

36. Holland, J.H. *Adaptation in Natural and Artificial Systems: An Introductory Analysis with Applications to Biology, Control, and Artificial Intelligence*; MIT Press: Cambridge, MA, USA, 1992.

37. Prins, C. A simple and effective evolutionary algorithm for the vehicle routing problem. *Comput. Oper. Res.* **2004**, *31*, 1985–2002. [CrossRef]

38. Moscato, P.; Cotta, C. An accelerated introduction to memetic algorithms. In *Handbook of Metaheuristics*; Springer: Boston, MA, USA, 2019; pp. 275–309.

39. Martínez-Salazar, I.A.; Molina, J.; Ángel-Bello, F.; Gómez, T.; Caballero, R. Solving a biobjective transportation location routing problem by metaheuristic algorithms. *Eur. J. Oper. Res.* **2014**, *234*, 25–36. [CrossRef]

40. Deb, K.; Pratap, A.; Agarwal, S.; Meyarivan, T. A fast and elitist multiobjective genetic algorithm: NSGA-II. *IEEE Trans. Evol. Comput.* **2002**, *6*, 182–197. [CrossRef]

41. Koulaeian, M.; Seidgar, H.; Kiani, M.; Fazlollahtabar, H. A Multi Depot Simultaneous Pickup and Delivery Problem with Balanced Allocation of Routes to Drivers. *Int. J. Ind. Eng. Theory Appl. Pract.* **2015**, *22*, 223–242.

42. Chunyu, R.; Xiaobo, W. Research on Multi-vehicle and Multi-Depot Vehicle Routing Problem with Time Windows for Electronic Commerce. In Proceedings of the 2010 International Conference on Artificial Intelligence and Computational Intelligence, Sanya, China, 23–24 October 2010; Volume 1, pp. 552–555, doi:10.1109/AICI.2010.121. [CrossRef]

43. Gillett, B.E.; Johnson, J.G. Multi-terminal vehicle-dispatch algorithm. *Omega* **1976**, *4*, 711–718. [CrossRef]

44. Augerat, P.; Belenguer, J.M.; Benavent, E.; Corberán, A.; Naddef, D.; Rinaldi, G. *Computational Results with a Branch and Cut Code for the Capacitated Vehicle Routing Problem*; Technical Report; IMAG, Institut National Polytechnique: Grenoble, France, 1995.

45. Taillard, É.D. A heuristic column generation method for the heterogeneous fleet VRP. *RAIRO-Oper. Res.* **1999**, *33*, 1–14. [CrossRef]

46. Li, F.; Golden, B.; Wasil, E. A record-to-record travel algorithm for solving the heterogeneous fleet vehicle routing problem. *Comput. Oper. Res.* **2007**, *34*, 2734–2742. [CrossRef]

47. Zitzler, E.; Laumanns, M.; Thiele, L. SPEA2: Improving the strength Pareto evolutionary algorithm. In *EUROGEN 2001, Evolutionary Methods for Design, Optimization and Control with Applications to Industrial Problems*; International Center for Numerical Methods in Engineering (CIMNE): Barcelona, Spain, 2002; pp. 95–100.

48. Zitzler, E.; Thiele, L. Multiobjective evolutionary algorithms: A comparative case study and the strength Pareto approach. *IEEE Trans. Evol. Comput.* **1999**, *3*, 257–271. [CrossRef]

applied sciences

Article

A Parallel Meta-Heuristic Approach to Reduce Vehicle Travel Time in Smart Cities

Hector Rico-Garcia [1], Jose-Luis Sanchez-Romero [1,*], Antonio Jimeno-Morenilla [1] and Hector Migallon-Gomis [2]

[1] Department of Computer Technology, University of Alicante, San Vicente del Raspeig, Alicante 03690, Spain; hector.rico@gmail.com (H.R.-G.); jimeno@dtic.ua.es (A.J.-M.)

[2] Department of Computer Engineering, Miguel Hernandez University, Elche, Alicante 03202, Spain; hmigallon@umh.es

* Correspondence: sanchez@dtic.ua.es

Abstract: The development of the smart city concept and inhabitants' need to reduce travel time, in addition to society's awareness of the importance of reducing fuel consumption and respecting the environment, have led to a new approach to the classic travelling salesman problem (TSP) applied to urban environments. This problem can be formulated as "Given a list of geographic points and the distances between each pair of points, what is the shortest possible route that visits each point and returns to the departure point?". At present, with the development of Internet of Things (IoT) devices and increased capabilities of sensors, a large amount of data and measurements are available, allowing researchers to model accurately the routes to choose. In this work, the aim is to provide a solution to the TSP in smart city environments using a modified version of the metaheuristic optimization algorithm Teacher Learner Based Optimization (TLBO). In addition, to improve performance, the solution is implemented by means of a parallel graphics processing unit (GPU) architecture, specifically a Compute Unified Device Architecture (CUDA) implementation.

Keywords: smart cities; meta-heuristics; travelling salesman problem; TLBO; parallelism; GPU

Citation: Rico-Garcia, H.; Sanchez-Romero, J.-L.; Jimeno-Morenilla, A.; Migallon-Gomis, H. A Parallel Meta-Heuristic Approach to Reduce Vehicle Travel Time in Smart Cities. *Appl. Sci.* **2021**, *11*, 818. https://doi.org/10.3390/app11020818

Received: 12 November 2020
Accepted: 13 January 2021
Published: 16 January 2021

Publisher's Note: MDPI stays neutral with regard to jurisdictional claims in published maps and institutional affiliations.

1. Introduction

The Smart City concept involves providing the urban environment with smart infrastructure, technology, and procedures to ameliorate the quality of life of inhabitants from an integrated perspective. Among the different aspects that have an impact on improving the quality of life, several action fronts can be noted related to the movement of people and goods, that is, traffic in the urban or metropolitan environment [1]. Indeed, effective traffic management results in a reduction in the travel times of citizens in their vehicles, which in turn has a positive impact on reducing the stress levels of drivers and optimizing the arrival at their workplaces for the effective performance of different professional activities. In addition, effective traffic management also entails better use of existing infrastructure and a reduction in pollution levels, an issue that is becoming increasingly relevant given people's awareness of environmental care and the impact on health.

The problem of traffic in cities becomes more relevant if we take into account the displacements that must be made by carriers and couriers to take different goods from headquarters to different delivery points, whether they are supermarkets, offices, private or homes. It should also be pointed out that the problem will affect the different platforms of autonomous vehicle fleets in the near future, in which the driver takes a passive role and the vehicle makes the decisions regarding the route to be followed to reach its destination in the shortest time.

The large amount of data and measurements that Internet of Things (IoT) devices can provide allows researchers to accurately model the different routes within an urban environment [2]. In this way, as explained in [3,4], smart cities should provide user-centered mobility services, implementing intelligent and/or automated transportation systems that incorporate strategies of artificial intelligence and technologies, such as the IoT and Physical Internet (PI). The integration of different technologies, such as RFID and LoRaWAN (Long-Range WAN), to implement PI oriented to provide Mobility-as-a-Service is demonstrated in [5].

With the information available, routes are not only determined by the distance between different geographical points, but can also incorporate aspects related to the expected time of arrival, which include, in addition to the distance itself, traffic lights and their crossing times, pedestrian crossings, and possible traffic jams.

A correspondence can be established between this problem and the classic combinatorial travelling salesman problem (TSP), in this case applied to car movements within urban environments. This problem can be informally described in the following manner: "Given a list of geographical points and the distances between each pair of points, what is the shortest possible path that passes one and only one point and returns to the starting point?".

The problem has given rise to a wide variety of research, and various algorithms and heuristics have been developed to try to reduce the complexity when obtaining solutions. In this paper, the aim is to provide an optimal or suboptimal solution to TSP instances applied to smart city environments, using a modified version of the metaheuristic optimization algorithm Teacher Learner Based Optimization (TLBO). In addition, performance is improved by implementing the solution on a parallel graphics processing unit (GPU) architecture, specifically a Compute Unified Device Architecture (CUDA) implementation.

This paper is organized as follows: Section 2 provides a review of state-of-the-art traffic congestion management in smart cities. Section 3 introduces a formal description of the TSP and its similarity with the problem of managing traffic in urban environments. Section 4 describes the TLBO method applied to both continuous and discrete domains. Section 5 shows the manner in which TLBO is implemented on a parallel CUDA architecture to enhance its performance. Section 6 depicts the results of the parallel TLBO when applied to a set of real scenarios from a well-known benchmark. Finally, in Section 7 conclusions of the research work are outlined and future lines of research are proposed.

2. The Concern of Traffic Management in Smart Cities and Urban Environments

One of the main problems to be addressed in an urban environment is vehicle traffic management. Although freight traffic is almost inevitably composed of trucks and vans, passenger movement throughout a city is still mostly operated by private vehicles [6]. As an example, 45% of American people had no access to public transportation in 2018 [7].

Numerous issues regarding the problems associated with car traffic in urban environments are currently being studied. Because the main objective in designing a smart city is to improve "personal satisfaction by utilizing innovation to enhance the proficiency of administrations and address occupants issues" [2], the movements of inhabitants through the city must be managed in such a way that travel time is reduced according to the specific needs of each inhabitant, but concerns about pollution and well-being must also be addressed. The same issues apply to freight transport, because transporters must make their deliveries in an efficient way, trying to reduce delivery times, pollution rates, and stress and tiredness of drivers.

Several research works can be found that deal with the possibility of monitoring traffic in urban environments by means of different technologies and methods. In the work of Rizwan et al. [2], an inexpensive real-time traffic management system is proposed to provide a smart service by activating traffic indicators to instantly update the traffic information. Inexpensive vehicle detection sensors are embedded in the middle of the road every 500 m or 1000 m; IoT is used to quickly acquire traffic information and send it for processing; and real-time flow data is sent for Big Data analysis. In the work of Fernandez-Ares [8], an approach was developed which tracks the movement of people and vehicles by monitoring the radioelectric space, capturing Wi-Fi and Bluetooth signals supplied by smartphones or on-board hands-free devices. In the work of Sendra et al. [9], a collaborative Long-Range (LoRa)-based sensor network to monitor the pollution levels in smart cities is developed. The system consists of geo-located nodes embedded in vehicles and fixed places to monitor temperature, relative humidity, and concentration of CO_2 in urban environments. The system uses the gathered data to generate the necessary orders to traffic signs and panels that control the traffic circulation. In Kazmi et al. [10], a methodology is proposed that uses VITAL-OS, an IoT platform that enables collection, integration, and management of data streams coming from multifunctional IoT devices and data sources. The information gathered is analyzed to detect traffic noise events; if the noise level on a particular road is higher than a certain threshold, the required traffic signal junction is managed for traffic re-routing to alternative paths. In Kök et al. [11], a deep learning model is proposed for analyzing IoT smart city data. The model is based on Long Short Term Memory (LSTM) networks to predict future values of air quality. In Jabbarpour et al. [12], an IoT application called Intelligent Guardrails is shown. It uses vehicular networks to identify traffic situation of the roads, and incorporates electronic and mechanical techniques to increase the number of lanes of the congested side of a highway while decreasing the lanes on the non-congested side. In Singh et al. [13], a visual Big Data analytics framework for automatic detection of bikers who do not wear a helmet in city traffic is proposed. The paper discusses the issues involved in visual Big Data analytics for traffic control in a city on a surveillance data scope. In Pawłowicz et al. [14], an approach of a traffic management system is proposed. It includes 5G communication, RFID-based parking space monitoring, and cloud services for supervision and machine learning. In Rathore et al. [15], a system for a smart digital city that uses IoT devices for collecting city data and Big Data analytics for acquiring knowledge is proposed. A model is proposed to develop a system that can handle a large volume of city data and provide guidelines to the local authorities. The system implementation involves several stages, including data generation and gathering, aggregation, filtering, classification, preprocessing, computing, and decision making. The gathered data from smart components within the city is processed in real-time to achieve a smart service using Hadoop under Apache Spark. Data generated by smart homes, smart parking areas, weather systems, pollution monitoring sensors, and vehicle networks are used for analysis and testing. In the work of Behnke and Kirschstein [16], a study on the effect of path selection in emission-oriented urban vehicle routing is presented, which is composed of a heterogeneous road network with regard to speed and acceleration frequency. An algorithm to determine all means of emission minimization for pairs of origin and destination nodes given a specific vehicle is proposed and tested using the road network from the city of Berlin.

In Ehmke et al. [17], two data-driven approaches for determining time-dependent emission minimizing paths in urban environments are proposed. Their performances are compared with respect to computing efficiency and solution quality. On average, emissions-minimizing methods can reduce emissions by approximately 3.5% compared to distance-minimizing paths, and 5.0% with regard to minimum time-dependent paths. The emissions-optimized path is roughly 4% longer than the paths generated by distance-based methods, and around 6% longer with regard to travel time compared to travel-time optimized paths.

In the work of Suzuki [18], attention is paid on the so-called pollution routing problem (PRP), which tries to minimize the fuel consumption or pollutants emission of trucks. The final goal is to develop a decision support system of pollution vehicle routing for eco-friendly enterprises. The work identifies, by analyzing the state-of-the-art PRP and gathering expert opinions from carrier managers, a practical PRP model that uses a minimal subset of the main factors affecting fuel consumption, and then develops a solution for this model. In Ehmke et al. [19], research is focused on the problem of minimizing CO_2 emissions in the routing of a fleet of capacitated vehicles in urban environments. A local search procedure, named the tabu search heuristic, is adapted to solve the problem. The research uses instances from a real road network dataset and 230 million speed observations. In Kramer et al. [20], a metaheuristic approach, called ILS-SOA-SP, is proposed to solve the PRP; this method integrates iterated local search (ILS) with a set partitioning (SP) procedure and a speed optimization algorithm (SOA). The proposed method has the benefit of performing combined route and speed optimization within several local search and integer programming components.

As a summary, a wide variety of research has been carried out with regard to the improvement of quality of life in smart cities in terms of traffic management and, consequently, the reduction of fuel consumption, street noise, and pollution.

3. The Travelling Salesman Problem

The TSP has given rise to a wide variety of research, due to its simplicity of description but its complexity at the time of obtaining a solution [21]. If the problem is formulated using graph theory terminology, it can be defined as a graph $G = (V, A)$, where $V = \{v_1, \ldots, v_n\}$ consists of a set of n vertices (nodes) and $A = \{(v_i, v_j)/v_i, v_j \in V, i \neq j\}$ is a set of edges with an associated non-negative cost (distance) matrix $D = (d_{ij})$. The problem is symmetric if $d_{ij} = d_{ji}$ for any pair of vertices $(v_i, v_j) \in A$, and it is said to be asymmetric (ATSP) otherwise.

If $I:\{1, \ldots, n\} \rightarrow \{1, \ldots, n\}$ is a bijective function that determines a reordering of vertices v_i in V, the TSP can be defined as obtaining the minimal result for the following calculation:

$$F_{TSP} = \sum_{j=1}^{n-1} d_{I(j)I(j+1)} \tag{1}$$

It is easy to establish an analogy or correspondence between the TSP and the problem of managing traffic in smart cities. For example, the set of nodes can be related to the different city locations where a van or truck must deliver its cargo from a depot. As previously explained, the meaning of *distance* in the case of urban environments is different from the mere concept of geographical distance, because it can also incorporate traffic lights and their crossing times, pedestrian crossings, speed limits, and even dynamic factors, such as temporary traffic jams, road maintenance works, and accidents. In addition, data collected on fuel consumption and pollution related to different roads can be added.

For a set of n nodes, obtaining the optimal solution by means of exhaustive search is a problem that implies $(n - 1)!$ comparisons. As an example, in the case of ten nodes, the required comparisons between possible solutions would be $9! = 362,880$, and 15 nodes would imply comparing $14! = 87,178,291,200$ possible solutions to obtain the best solution. The extensive list of research papers related to TSP has produced different methods that try to decrease the aforementioned complexity. Heuristic methods are suitable for managing TSP, and much attention has been paid to these types of methods among researchers on optimization. In [22], a review of swarm intelligence applied to graph search problems is contributed, but the paper focuses almost solely on ant colony optimization (ACO) and bee colony optimization (BCO).

In [23], two different modifications of the artificial bee colony (ABC) method are applied to TSP: a combinatorial algorithm, called CABC, and an improved version of CABC, called qCABC. Experiments are performed on a benchmark of 15 TSP instances. The performance of these two algorithms and eight different genetic algorithm (GA) versions is compared. Moreover, the performance of ABC is also compared with the ant colony system (ACS) and bee colony optimization methods. Results show the suitability of CBAC and qCABC algorithms to be applied to TSP problems.

Most of the works related to solving the TSP by means of metaheuristic optimization methods use the ant colony optimization method [24–28], sometimes in a hybrid version with other methods [29]. This is because the natural behavior of ants in the development of a colony can be easily matched to the problem of finding the shortest path between a series of nodes.

4. The TLBO Optimization Method

4.1. Original TLBO Formulation

Rao [30] developed a series of metaheuristic algorithms that do not require the adjustment of initial parameters to operate. Among these algorithms, TLBO imitates the learning process that takes place in the classroom: from the teacher to the students. It is an iterative process that evolves in stages to optimize a function with multiple variables. This process culminates when the conditions that make the solution valid are met.

For each stage, a random population of possible solutions (called individuals) is created. These values are assigned taking into account the domain of the function f. Once these values are assigned for the population, two stages are carried out, as detailed below.

4.1.1. Teacher Stage

Individual i within the population is represented by vector $X(i)$. Therefore, $X(i,j)$ refers to variable j of individual i. The teacher stage aims to assess the function for each member of the population considering its own parameters, $f(X(i))$. The individual with a value closest to the optimum is selected as the teacher (X_{best}). Then, teacher values are used to bring his disciples closer to him by following Equation (2), which also takes into account the mean values of the population. Thus, each individual will have its new (X_{new}) solution.

$$X_{new}(i,j) = X(i,j) + rand(0,1)(X_{best}(j) - TFactor \cdot X_m(j)) \tag{2}$$

As can be observed in Equation (2), the value of $X(i,j)$ is modified according to the distance that exists between the teacher and the mean of the population (X_m). The factor T is a random integer value that can be valued at 1 or 2, and whose calculation is shown in Equation (3). After changing the values of each individual, the function is evaluated again ($f(X_{new}(i))$). Only in the case that the evaluation offers a more optimal solution than the original one, the new values calculated for the individual i are taken.

$$TFactor = round(1 + rand(0,1)) \tag{3}$$

4.1.2. Learner Stage

At the learner stage, again all individuals are evaluated and compared. This time the comparison takes place with another randomly selected student. In this pair-wise comparison, the student with the optimal solution is called *Best Learner* and the other is called *Worst Learner*. Both students are used to set up a possible new individual X_{new} from the original $X(i)$ according to Equation (4). In case this proposal is more optimal than the original one, the new values calculated will replace the original ones for that individual.

$$X_{new}(i,j) = X(i,j) + rand(0,1) \cdot (BestLearner(j) - WorstLearner(j)) \tag{4}$$

In this algorithm only the number of iterations and the number of individuals in the population have to be fixed; therefore, its main advantage is that there are no parameters that have to be adjusted for the algorithm to converge towards an optimal solution, as happens in other metaheuristic algorithms. In the recent years, there has been an increase in the number of publications that highlight the advantages of TLBO for the resolution of a large number of engineering problems [31–34].

4.2. Discrete TLBO

The problem to be solved in this research, the TSP, is of a discrete nature which comes from the appropriate selection of a combination of sites whose sorting satisfies a criterion. In the case of this research, the aim is to find the shortest route.

This problem does not fit with the characteristics of the original TLBO, whose domain of solutions is established in a continuous range in which the selection of values taken from a set is not possible. For this reason, it is necessary to make a change in the original TLBO algorithm, so that discrete values taken from a set can be selected. This discrete version of the algorithm (DTLBO) is based on the research proposed in [35] and substantial changes have been made to adapt it to the TSP case study.

In [35], the whole population consists of 100 individuals. Although the paper does not mention it implicitly, it can be guessed that each subpopulation contains four individuals, so 25 subpopulations are supposed to be formed. The division of a population into different subpopulations is a strategy used in several metaheuristic methods to increase the diversity of the population and, therefore, reduce the probability of being trapped in a local optimum.

4.2.1. Representation of Individuals

The representation of individuals in DTLBO changes substantially from the original algorithm. In this version of the algorithm the individuals represent routes through a set of places. Each variable within an individual corresponds to a node or location. As shown in Figure 1, if a set of eight places has to be visited ($v_0, v_1 \ldots v_7$), an individual could be represented as eight connected nodes that determine the next order of visit: $v_7{\rightarrow}v_0{\rightarrow}v_1{\rightarrow}v_6{\rightarrow}v_4{\rightarrow}v_2{\rightarrow}v_5{\rightarrow}v_3$.

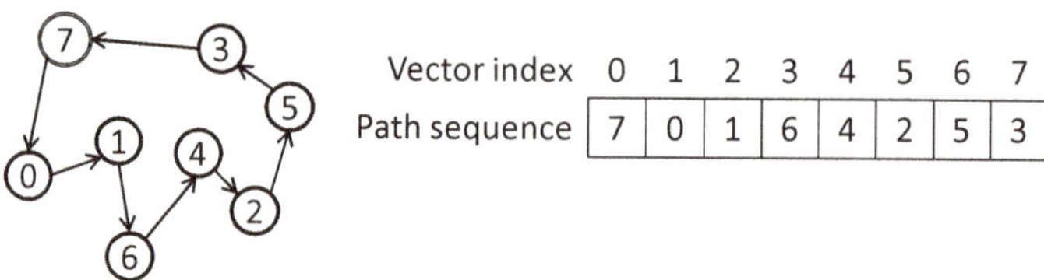

Figure 1. Representation of an individual solution.

In contrast to the work proposed in [35], which uses four completely random subpopulations, in this version of the DTLBO one of the individuals is not generated randomly but is assigned initial values based on a greedy strategy. This is intended to accelerate the convergence of the algorithm and preserve its ability to avoid local minima by maintaining randomness in the subpopulations.

4.2.2. DTLBO Teacher Stage

At this stage, a Partial Teacher is selected for each subpopulation. As with the original TLBO, this teacher is used to improve the students in each subpopulation taking into account the mean value of the subpopulation. In addition, to avoid losing the global vision of the route, a global Teacher is also selected as the best individual of the whole population. It must be taken into account that the calculation of the mean individual could violate the path conditions of the TSP, so a viability condition must be added to its calculation. This condition is performed in the following four steps

Step 1: The number of times a place appears is counted. To store the places that appear more than once, the *TempA* vector is used. Each place is stored in the last position (higher index) that it occupies in the original unfeasible vector; in case a place does not appear or just appears once, the corresponding item remains empty (symbol -). In Figure 2, it can be observed that locations 3 and 5 appear in the last positions they occupied in the original unfeasible vector, that is, positions 5 and 6, respectively.

Step 2: Search for the missing locations. Again, in Figure 2 it can be seen that places 1, 6, and 7 do not appear. The *TempB* vector is used to store these places. In this vector, the index of the place that does not appear is written in the associated item, leaving the remainder of the items empty.

Step 3: A vector *TempC* is created to indicate the places that appear just once. In Figure 2, it can be observed that places 0, 2, and 4 appear in their corresponding positions within *TempC*.

Step 4: Using the above mentioned vectors, a viable individual is constructed by adding to the empty items of *TempA* and *TempC* the places that do not appear in them and that are found in *TempB*.

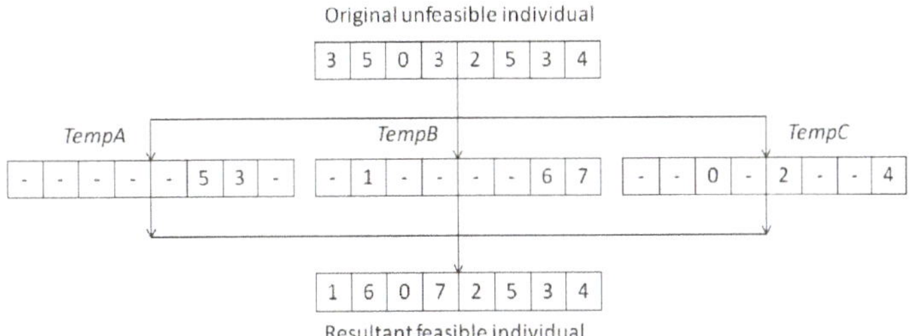

Figure 2. An example of the viability operation.

The crossover operation is used in the DTLBO to generate new individuals from existing ones. This operation is indicated by the symbol $\otimes$. In this version of the algorithm, four different ways of creating new individuals from crossovers were introduced and are shown in Equation (5). In contrast to the work presented in [35], in which a fixed cross is assigned between individuals, this approach makes a random selection for each cross operation. This randomness avoids falling into local minima and increases the variability of populations.

$$\begin{aligned} Xnew(i) &= X(i) \otimes Teacher \\ Xnew(i) &= X(i) \otimes PartialTeacher(i) \\ Xnew(i) &= X(i) \otimes Mean(i) \\ Xnew(i) &= PartialTeacher(i) \otimes Mean(i) \end{aligned} \tag{5}$$

The operation of the crossover can be shown in Figure 3. Suppose that two individuals A and B must generate a new one called A_c. The initial and final positions of the new individual are selected randomly and thus the new individual A_c replaces the selected items by those of individual B. Logically, after this crossover is performed, the viability operation must be applied.

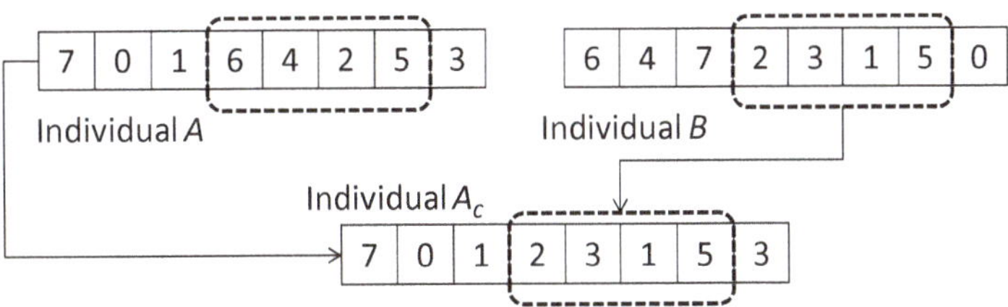

Figure 3. An example of the crossover operation.

To increase the variability of the populations, a mutation operator is included, which is denoted by the symbol Θ (shown in the Equation (6)). Figure 4 shows the functionality of this operator. Given an A_c individual, initial and final positions are randomly selected; in the current example, these are positions 3 and 6. Once these positions are determined, the elements included in this sequence (items of positions 3, 4, 5, and 6) are inverted to create the mutated element A_{cm}.

$$A_{cm}(i) = \Theta A_c(i) \tag{6}$$

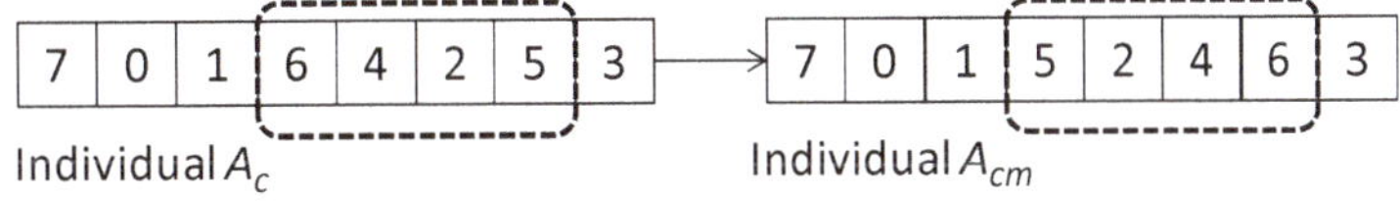

Figure 4. An example of the mutation operation.

4.2.3. DTLBO Learner Stage

Similar to the original TLBO, in the learner stage a random individual k is chosen for each individual i in the subpopulation to be compared with. The new individual is created from Equation (7), where $\otimes$ represents the crossover operation, working in the same way as the crossover operation in the teacher stage. Then it will be necessary to apply the condition of viability on the new individual $X_{new}(i)$ and then apply the mutation operator as described in the teacher stage.

$$Xnew(i) = X(i) \otimes X(k) \tag{7}$$

Although the accuracy of the DTLBO presented in [35] is best when compared with well-known algorithms such as ACO, Particle Swarm Optimization (PSO), and Genetic Algorithm (GA), and only in some specific cases it is slightly beaten by ACO and ABC. It is also the case that its performance is low when cities have a large number of places to visit. For this reason, this investigation focused on a revision of this algorithm based on the improvement of its convergence and on a parallel implementation that allows taking advantage of the performance of the multiprocessing present in the existing GPUs.

5. Parallel TLBO Implementation on GPU

As explained above, applying a series of modifications to the TLBO algorithm can achieve significant results in discrete combinatorial problems, in this case, in the TSP problem. Satisfying results are achieved compared to other algorithms [35].

The changes implemented in the original TLBO cause the different stages to be more complex and add a higher computational cost to the algorithm. As a consequence, the iterations of the DTLBO are significantly more expensive than those of the original TLBO, and therefore the computation time is penalized. With the aim of minimizing the impact of these modifications and the extra cost of computing on the performance of the algorithm, a parallel implementation of the algorithm was developed using a CUDA architecture in a GPU environment. Although parallelization of an algorithm using CUDA is not always the best solution when applying parallelization techniques, the specific features of DTLBO can be exploited to provide a remarkable improvement in performance when compared to sequential solutions using this parallel architecture. Previous research has attempted improvement of metaheuristic methods applied to graph search problems by means of parallel implementations [36].

The initial and fundamental step when proceeding with the algorithm parallelization consists in creating an adequate design of the memory structure and the execution flow to minimize the global thread blockages. If GPU memory is mismanaged, the impact in execution time can be detrimental because transference operations between the different memory levels within the GPU imply a high latency. Therefore, these transferences should be minimized.

5.1. Design of the Memory Organization

The first task that must be carried out to parallelize the algorithm is to conceive an adequate design of the memory structure and the execution flow to minimize the blockage of the different threads executing in parallel within the GPU. Thus, the different GPU memory levels are organized to manage different kinds of data (thread, subpopulation, and global levels) as shown in Figure 5.

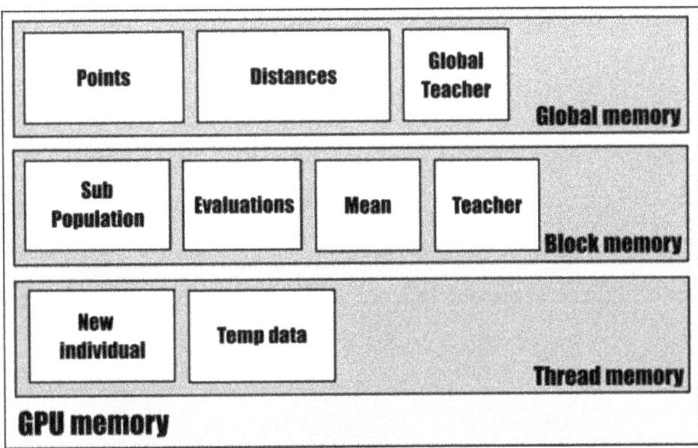

Figure 5. Organization of the Compute Unified Device Architecture (CUDA) memory.

Global memory. This memory stores two main data blocks: an array that contains the required points for generating the individuals of the populations in each block; and an array that contains the whole set of pre-calculated distances between the different points. Global memory also stores the best individual of the whole population and its evaluation (global Teacher), which are globally calculated in each iteration to be used by every individual within the population. Much of the data stored in the global memory will be only read but never modified, and will be used to avoid computing the total distance travelled when evaluating each individual. Only the best evaluation (global Teacher) must be modified if required each iteration.

Block memory. This memory stores the data shared by a subpopulation. The memory is organized in a matrix; each row corresponds to an individual, and each column stores the index of an urban node. An extra column is in charge of storing the individual's solution to avoid repeating the evaluation.

Thread memory. This consists of a private, local memory to each thread which is not shared with the remainder of individuals. The variables stored in this memory are used for the calculations needed at each stage of the DTLBO, and are updated during each of the iterations.

5.2. Execution Flow

An execution flow was devised to minimize the blockages of the threads at the time of synchronization when running the algorithm (*syncthreads*). The TLBO consists of a series of phases linked in such a way that they make it necessary to synchronize the threads to obtain common data for the whole population, such as the mean individual and the best individual (Teacher). In addition, to obtain this information, the reduction technique is utilized to achieve a minimal number of iterations needed to obtain the aforementioned values from the population. To build a parallel execution of the algorithm, each thread corresponds to an individual of the population. The first thread (with index 0) of each subpopulation is responsible for carrying out the operations in the local memory, and the first thread of the first population is responsible for carrying out the necessary operations in the global memory. This is represented in Figure 6. In this manner, conflicts and delays in the access to the different memory levels are avoided.

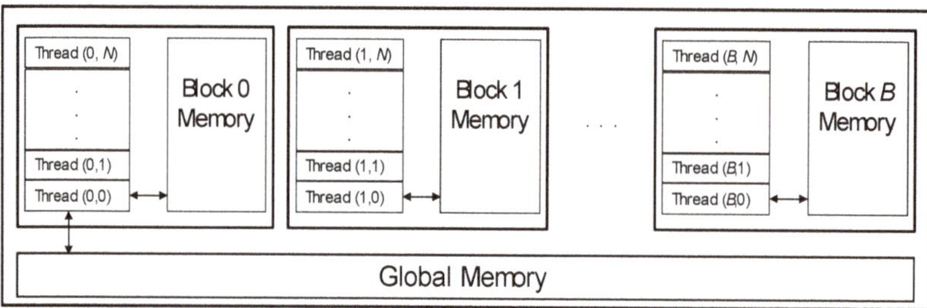

Figure 6. Subpopulation organization and memory accesses.

6. Experimentation

Experimentation was performed to compare the execution time when solving four TSP problems from the TSPLIB library [37]. A sequential implementation and a manycore GPU parallel implementation based on CUDA (9.2 version) were compared. The system used for performing the experiments was equipped with a Pentium i7 3.2 GHz and an NVIDIA GeForce 1060p graphics accelerator, 6 GB GPU RAM, and 32 GB DDR4 RAM. Different scopes were devised for each of the problems by varying the population sizes (64 and 128 individuals) and the number of iterations performed (1000, 5000, and 10,000 iterations). The cities from TSPLIB that were considered for the experiments were Berlin52, Att48, Eil76, and Ch130, with 52, 48, 76, and 130 urban nodes, respectively. They represent real scenarios taken from cities in Europe, USA, and China.

Figures 7–14 depict the results with respect to the CPU and GPU mean time in milliseconds (ms) of 10 blocks of 1000 runs, each with a PC clean reset. Results show that the DTLBO GPU parallel version improves performance up to 6× when dealing with a high number of individuals and iterations. DTLBO reached the optimal path in Att48 (distance = 33,523); in the case of Berlin52, the difference obtained with regard to the optimal was only 2 (7544 versus 7542); in the case of the problems with a higher number of urban nodes, DTBLO obtained 6336 in Ch130 (optimal = 6110), and 552.63 in Eil76 (optimal = 538). The paths obtained for these four real urban scenarios are shown in Figures 15–18.

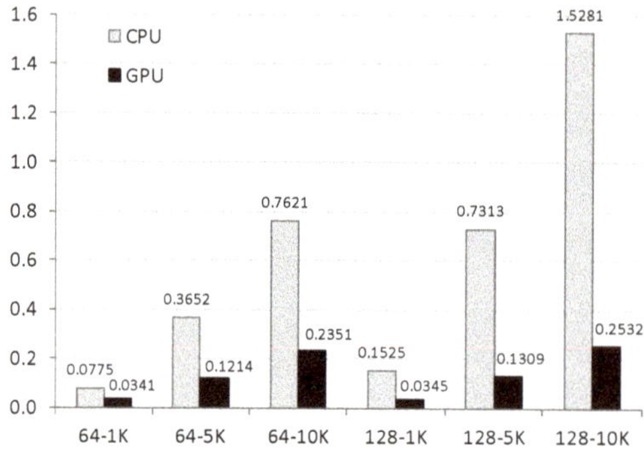

Figure 7. CPU and graphics processing unit (GPU) times (in ms) from Berlin130 with respect to different populations (64 and 128) and iterations (1000, 5000, and 10,000).

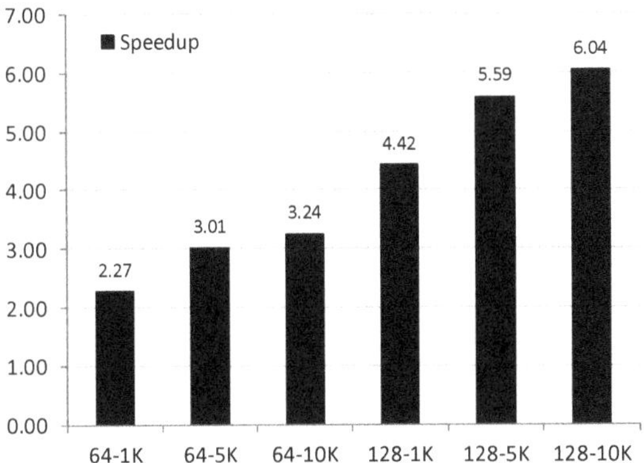

Figure 8. Speedup when comparing CPU and GPU times from Berlin130 with respect to different populations (64 and 128) and iterations (1000, 5000, and 10,000).

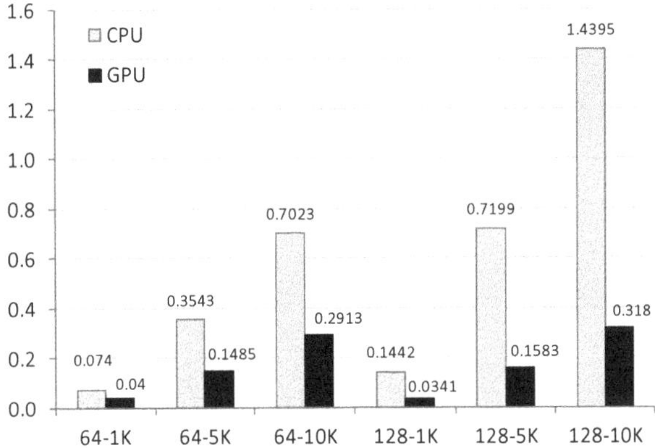

Figure 9. CPU and GPU times (in ms) for Att48 with respect to different populations (64 and 128) and iterations (1000, 5000, and 10,000).

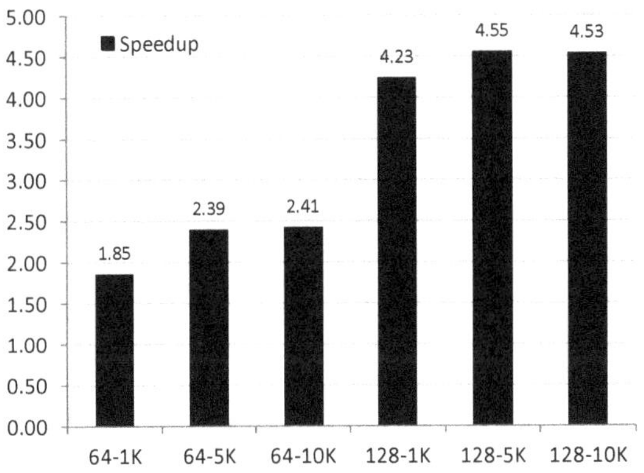

Figure 10. Speedup when comparing CPU and GPU times for Att48 with respect to different populations (64 and 128) and iterations (1000, 5000, and 10,000).

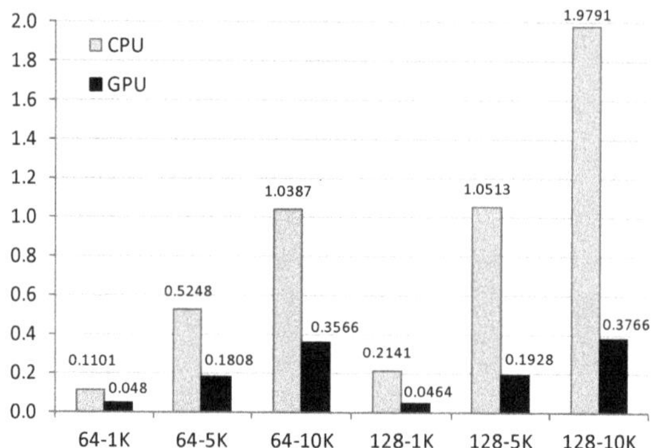

Figure 11. CPU and GPU times for Eil76 (in ms) with respect to different populations (64 and 128) and iterations (1000, 5000, and 10,000).

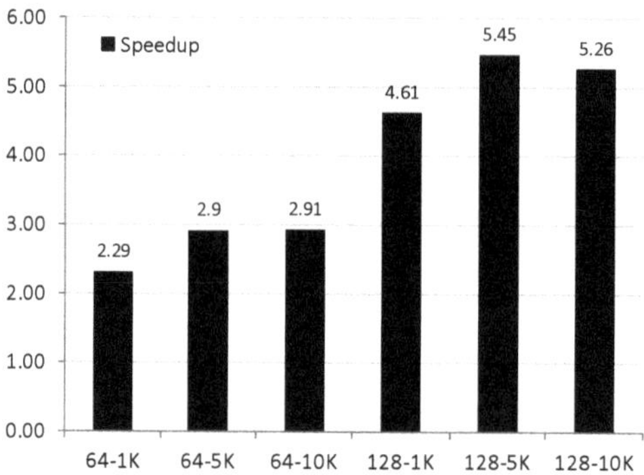

Figure 12. Speedup when comparing CPU and GPU times for Eil76 with respect to different populations (64 and 128) and iterations (1000, 5000, and 10,000).

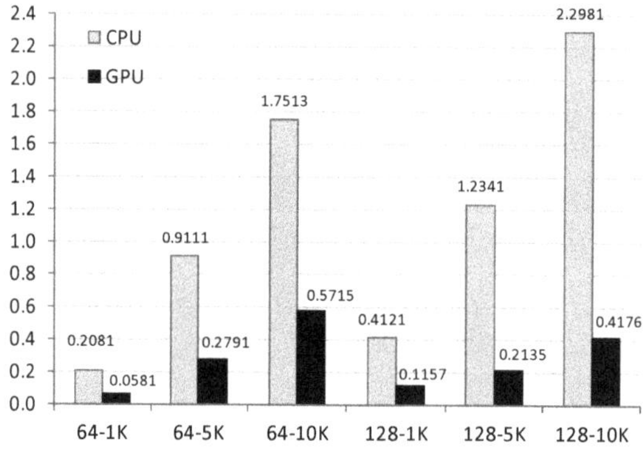

Figure 13. CPU and GPU times (in ms) for Ch130 with respect to different populations (64 and 128) and iterations (1000, 5000, and 10,000).

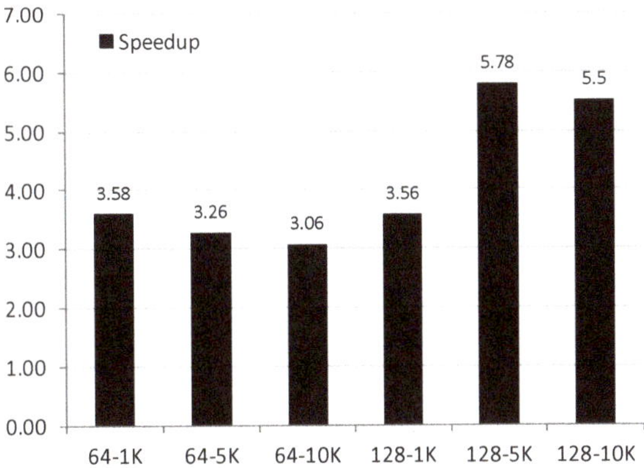

Figure 14. Speedup when comparing CPU and GPU times for Ch130 with respect to different populations (64 and 128) and iterations (1000, 5000, and 10,000).

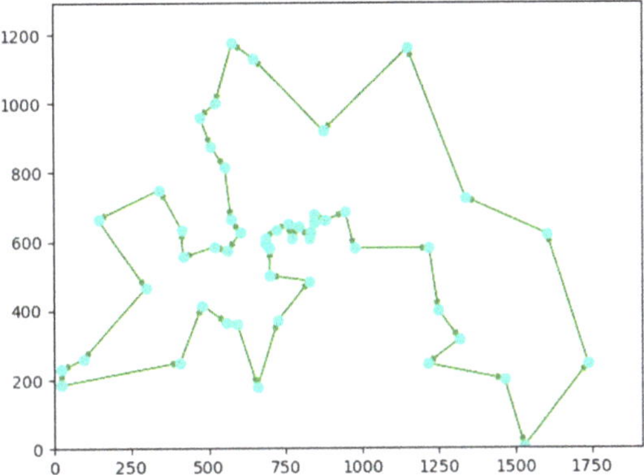

Figure 15. DTLBO solution for Berlin52.

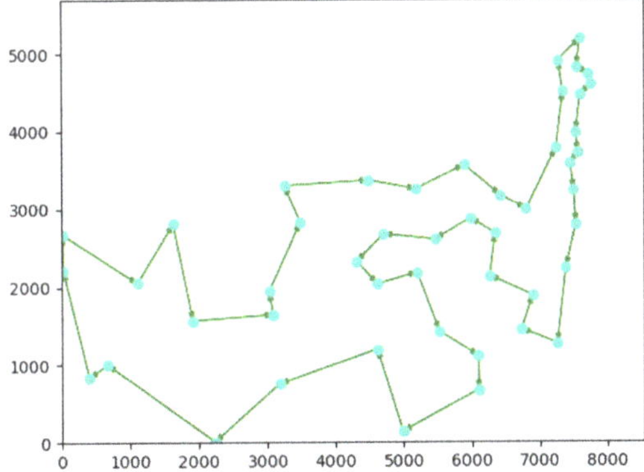

Figure 16. DTLBO solution for Att48.

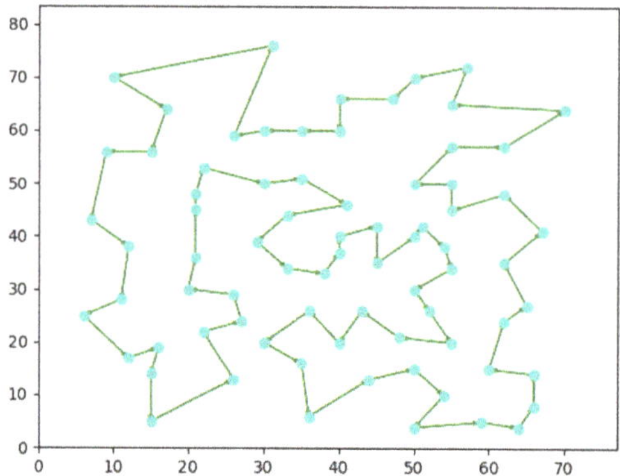

Figure 17. DTLBO solution for Eil76.

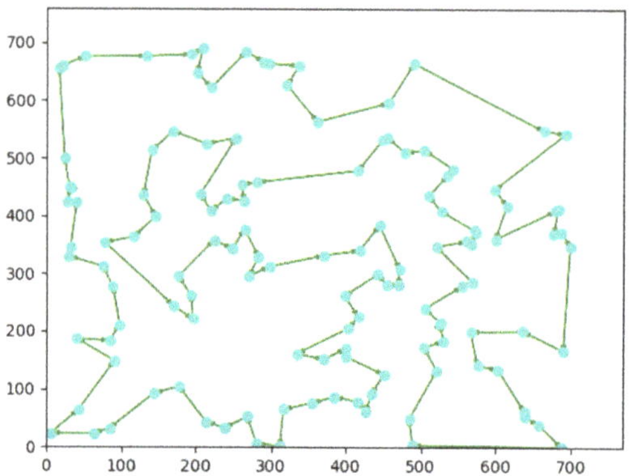

Figure 18. DTLBO solution for Ch130.

The GPU's features allow it to process a high number of threads per block. As a consequence, it was noted that the increase in the population does not have as much time penalty as the CPU, so the GPU times are only affected by the number of iterations. Moreover, the GPU block architecture is highly suitable for subpopulation algorithms, due to the fact that they can be placed in different blocks of the GPU and consequently run in parallel if the number of subpopulations allows it. It is possible for different problems to be executed on the same GPU, achieving a maximal utilization of the GPU resources and a minimal response time. All of the optimal solutions were obtained with populations of 128 individuals. With populations of 64 individuals, it was only possible to generate one of the optimal solutions (the one of the smallest problem).

It was established that population sizes of 64 and 128 are capable of computing all scenarios without memory problems: in the case of 64 individuals, to have a population size close to the number of variables of some problems; and in the case of 128 individuals, to have a population size above almost all of the problems. The maximum of 128 is also related to the limits of memory per block. The data stored at block level to be shared among the whole population for learning, and the comparisons between individuals, limit the population size. Although there are no thread-level memory conflicts, the information needed to make certain calculations, especially for the calculation of valid solutions and the choice of learners, makes it possible for memory conflicts to arise in some of the scenarios.

7. Conclusions

In this research, a parallel implementation of the discrete Teaching Learning Based Optimization algorithm (DTLBO) using a manycore GPU environment to improve the algorithm performance was proposed and applied to provide optimal or suboptimal solutions to the traveling salesman problem. Previous research has found this algorithm to be an excellent option for solving the TSP, but its biggest drawback is its complexity and computational cost when compared to other metaheuristic optimization algorithms. The parallel implementation using a manycore GPU proposed and developed in this research work substantially improved the algorithm performance, achieving important speedups with respect to a sequential implementation. When a high number of individuals and iterations are considered, the speedup is high as $6\times$. The results show that the algorithm is adequate to be applied to problems in which the execution time is one of the determining factors, especially when there are numerous urban nodes in the route that must be traveled in an optimal way.

To summarize, the original DTLBO method was modified by including a greedy strategy when searching for a first initial best individual instead of a purely random generation. When creating new individuals from crossovers, four different ways were considered, with one of them randomly selected each time. Indeed, because the method is time-consuming, another significant contribution of the work consists in accelerating the algorithm by means of a parallel implementation supported by CUDA architecture. The parallelization of this kind of algorithm is not a trivial task and, in fact, the results obtained when parallelizing an algorithm are not always better than those obtained with a sequential implementation. The parallel implementation of DTLBO proposed in this research work achieves satisfactory speedup results.

The developed implementation could be further enhanced in several parts by focusing on improvements in thread block management to optimize the use of GPU resources.

Moreover, although in this research work parallel implementations of the algorithm on CUDA were performed on a desktop computer, these algorithms can be executed in embedded systems inside cars, because some cars already exist with NVIDIA chipsets and support for CUDA, that are used for image recognition within smart driving. With these systems, smart driving could be improved by performing route enhancement using checkpoints within the GPU processing without penalizing the CPU. This is an important issue due to the fact that, in embedded systems in cars, the CPU is a highly demanded resource. Thus, it can be troublesome to run intensive processes on the CPU because the CPU must manage other subsystems in the vehicle. Another line of future research is the adoption of a multi-objective strategy aimed at improving travel time, which would include factors other than distance, such as traffic lights, car accidents, and road works.

Author Contributions: H.R.-G.: Conceptualization, Methodology, Software, Validation, Investigation, writing—original draft, Writing—review & editing. J.-L.S.-R.: Conceptualization, Methodology, Software, Validation, Investigation, writing—original draft, Writing—review & editing, Supervision, Project administration. A.J.-M.: Validation, Investigation, writing—original draft, Writing—review & editing, Supervision, Project administration, Funding acquisition. H.M.-G.: Validation, Investigation, writing—original draft, Writing—review & editing, Supervision, Funding acquisition. All authors have read and agreed to the published version of the manuscript.

Funding: This research was supported by the Spanish Ministry of Science, Innovation and Universities and the Research State Agency under Grant RTI2018-098156-B-C54 co-financed by FEDER funds, and by the Spanish Ministry of Economy and Competitiveness under Grant TIN2017-89266-R, co-financed by FEDER funds.

Institutional Review Board Statement: Not applicable.

Informed Consent Statement: Not applicable.

Data Availability Statement: Not applicable.

Conflicts of Interest: The authors declare no conflict of interest.

References

1. Benevolo, C.; Dameri, R.P.; D'Auria, B. Smart Mobility in Smart City. In *Empowering Organizations, Lecture Notes in Information Systems and Organisation*; Torre, T., Braccini, A., Spinelli, R., Eds.; Springer: Cham, Switzerland, 2016; Volume 11.
2. Rizwan, P.; Suresh, K.; Babu, M.R. Real-Time Smart Traffic Management System for Smart Cities by using Internet of Things and Big Data. In Proceedings of the 2016 International Conference on Emerging Technological Trends (ICETT), New York, NY, USA, 1–6 August 2016.
3. Nikitas, A.; Michalakopoulou, K.; Njoya, E.T.; Karampatzakis, D. Artificial Intelligence, Transport and the Smart City: Definitions and Dimensions of a New Mobility Era. *Sustainability* **2020**, *12*, 2789. [CrossRef]
4. Cruz, C.O.; Sarmento, J.M. 'Mobility as a Service' Platforms: A Critical Path towards Increasing the Sustainability of Transportation Systems. *Sustainability* **2020**, *12*, 6368. [CrossRef]
5. Karampatzakis, D.; Avramidis, G.; Kiratsa, P.; Tseklidis, I.; Oikonomidis, C. A Smart Cargo Bike for the Physical Internet enabled by RFID and LoRaWAN. In Proceedings of the 2019 Panhellenic Conference on Electronics & Telecommunications (PACET), University of Thessaly, Volos, Greece, 8–9 November 2019.
6. Department for Transport. Transport Statistics Great Britain: 2019 Summary. Available online: https://assets.publishing.service.gov.uk/government/uploads/system/uploads/attachment_data/file/870647/tsgb-2019.pdf (accessed on 8 April 2020).
7. American Public Transportation Association. Public Transportation Facts. Available online: https://www.apta.com/news-publications/public-transportation-facts (accessed on 8 April 2020).
8. Fernández-Ares, A.; Mora, A.; Arenas, M.G.; García-Sanchez, P.; Romero, G.; Rivas, V.; Castillo, P.; Merelo, J. Studying real traffic and mobility scenarios for a Smart City using a new monitoring and tracking system. *Futur. Gener. Comput. Syst.* **2017**, *76*, 163–179. [CrossRef]
9. Sendra, S.; Garcia-Navas, J.L.; Romero-Diaz, P.; Lloret, J. Collaborative LoRa-Based Sensor Network for Pollution Monitoring in Smart Cities. In Proceedings of the 2019 Fourth International Conference on Fog and Mobile Edge Computing (FMEC), Rome, Italy, 10–13 June 2019.
10. Kazmi, A.; Tragos, E.; Serrano, M. Underpinning IoT for road traffic noise management in smart cities. In Proceedings of the 2018 IEEE International Conference on Pervasive Computing and Communications Workshops (PerCom Workshops), Athens, Greece, 19–23 March 2018.
11. Kök, İ.; Şimşek, M.U.; Özdemir, S. A deep learning model for air quality prediction in smart cities. In Proceedings of the 2017 IEEE International Conference on Big Data (Big Data), Boston, MA, USA, 11–14 December 2017.
12. Jabbarpour, M.R.; Nabaei, A.; Zarrabi, H. Intelligent Guardrails: An IoT application for vehicle traffic congestion reduction in smart city. In Proceedings of the 2016 IEEE International Conference on Internet of Things (Ithings) and IEEE Green computing and communications (Greencom) and IEEE Cyber, Physical and Social Computing (cpscom) and IEEE Smart Data (smartdata), Chengdu, China, 15–18 December 2016.
13. Singh, D.; Vishnu, C.; Mohan, C.K. Visual Big Data analytics for traffic monitoring in smart city. In Proceedings of the 2016 15th IEEE international conference on machine learning and applications (ICMLA), Anaheim, CA, USA, 18–20 December 2016.
14. Pawłowicz, B.; Salach, M.; Trybus, B. Smart city traffic monitoring system based on 5G cellular network, RFID and machine learning. In Proceedings of the KKIO Software Engineering Conference, Pultusk, Poland, 27–28 September 2018; pp. 151–165.
15. Rathore, M.M.; Paul, A.; Hong, W.-H.; Seo, H.; Awan, I.; Saeed, S. Exploiting IoT and big data analytics: Defining Smart Digital City using real-time urban data. *Sustain. Cities Soc.* **2018**, *40*, 600–610. [CrossRef]
16. Behnke, M.; Kirschstein, T. The impact of path selection on GHG emissions in city logistics. *Transp. Res. Part E Logist. Transp. Rev.* **2017**, *106*, 320–336. [CrossRef]
17. Ehmke, J.F.; Campbell, A.M.; Thomas, B.W. Data-driven approaches for emissions-minimized paths in urban areas. *Comput. Oper. Res.* **2016**, *67*, 34–47. [CrossRef]
18. Suzuki, Y. A. dual-objective metaheuristic approach to solve practical pollution routing problem. *Int. J. Prod. Econ.* **2016**, *176*, 143–153. [CrossRef]
19. Ehmke, J.F.; Campbell, A.M.; Thomas, B.W. Vehicle routing to minimize time-dependent emissions in urban areas. *Eur. J. Oper. Res.* **2016**, *251*, 478–494. [CrossRef]
20. Kramer, R.; Subramanian, A.; Vidal, T.; Cabral, L.D.A.F. A matheuristic approach for the Pollution-Routing Problem. *Eur. J. Oper. Res.* **2015**, *243*, 523–539. [CrossRef]
21. Rego, C.; Gamboa, D.; Glover, F.; Osterman, C. Traveling salesman problem heuristics: Leading methods, implementations and latest advances. *Eur. J. Oper. Res.* **2011**, *211*, 427–441. [CrossRef]
22. Ilie, S.V. Survey on distributed approaches to swarm intelligence for graph search problems. *Ann. Univ. Craiova-Math. Comput. Sci. Ser.* **2014**, *41*, 251–270.
23. Karaboga, D.; Gorkemli, B. Solving Traveling Salesman Problem by Using Combinatorial Artificial Bee Colony Algorithms. *Int. J. Artif. Intell. Tools* **2019**, *28*, 1950004. [CrossRef]
24. Jabir, E.; Panicker, V.V.; Sridharan, R. Design and development of a hybrid ant colony-variable neighbourhood search algorithm for a multi-depot green vehicle routing problem. *Transp. Res. Part D Transp. Envi.* **2017**, *57*, 422–457. [CrossRef]
25. Gan, R.; Guo, Q.; Chang, H.; Yi, Y. Improved ant colony optimization algorithm for the traveling salesman problems. *J. Syst. Eng. Electron.* **2010**, *21*, 329–333. [CrossRef]

26. Shokouhifar, M.; Sabet, S. PMACO: A pheromone-mutation based ant colony optimization for traveling salesman problem. In Proceedings of the 2012 International Symposium on Innovations in Intelligent Systems and Applications, Trabzon, Turkey, 2–4 July 2012.

27. Bai, J.; Yang, G.; Chen, Y.-W.; Hu, L.-S.; Pan, C.-C. A model induced max-min ant colony optimization for asymmetric traveling salesman problem. *Appl. Soft Comput.* **2013**, *13*, 1365–1375. [CrossRef]

28. Głabowski, M.; Musznicki, B.; Nowak, P.; Zwierzykowski, P. Shortest Path Problem Solving Based on Ant Colony Optimization Metaheuristic. *Image Process. Commun.* **2012**, *17*, 7–17. [CrossRef]

29. Mahi, M.; Baykan, Ö.K.; Kodaz, H. A new hybrid method based on Particle Swarm Optimization, Ant Colony Optimization and 3-Opt algorithms for Traveling Salesman Problem. *Appl. Soft Comput.* **2015**, *30*, 484–490. [CrossRef]

30. Rao, R.V.; Savsani, V.J.; Vakharia, D.P. Teaching–learning-based optimization: A novel method for constrained mechanical design optimization problems. *Comput. Des.* **2011**, *43*, 303–315. [CrossRef]

31. Rao, R.V.; Patel, V. An elitist teaching-learning-based optimization algorithm for solving complex constrained optimization problems. *Int. J. Ind. Eng. Comput.* **2012**, *3*, 535–560. [CrossRef]

32. Rao, R.V.; Patel, V. Comparative Performance of an elitist Teaching-Learning-Based Optimization algorithm for solving unconstrained optimization problems. *Int. J. Ind. Eng. Comput.* **2013**, *4*, 29–50. [CrossRef]

33. Ebraheem, M.; Jyothsna, T.R. Comparative performance evaluation of Teaching Learning Based Optimization against genetic algorithm on benchmark functions. In Proceedings of the 2015 Power, Communication and Information Technology Conference (PCITC), Bhubaneswar, India, 15–17 October 2015; pp. 327–331.

34. Shah, S.R.; Takmare, S.B. A Review of Methodologies of TLBO Algorithm to Test the Performance of Benchmark Functions. *Program. Device Circuits Syst.* **2017**, *9*, 141–145.

35. Wu, L.; Zoua, F.; Chen, D. Discrete Teaching-Learning-Based Optimization Algorithm for Traveling Salesman Problems. In Proceedings of the MATEC Web of Conferences 128, 02022 EDP Sciences (2017), Zhuhai, China, 23–24 September 2017.

36. Arnautovic, M.; Curic, M.; Dolamic, E.; Nosovic, N. Parallelization of the ant colony optimization for the shortest path problem using OpenMP and CUDA. In Proceedings of the 2013 36th International Convention on Information and Communication Technology, Electronics and Microelectronics (MIPRO), Opatija, Croatia, 20–24 May 2013.

37. Universität Heidelberg. Institut für Informatik. TSPLIB. Available online: http://comopt.ifi.uni-heidelberg.de/software/TSPLIB95/ (accessed on 8 April 2020).

Article

Thread-Aware Mechanism to Enhance Inter-Node Load Balancing for Multithreaded Applications on NUMA Systems

Mei-Ling Chiang * and Wei-Lun Su

Department of Information Management, National Chi Nan University, Puli 54516, Taiwan; s94213034@gmail.com
* Correspondence: joanna@mail.ncnu.edu.tw

Abstract: NUMA multi-core systems divide system resources into several nodes. When an imbalance in the load between cores occurs, the kernel scheduler's load balancing mechanism then migrates threads between cores or across NUMA nodes. Remote memory access is required for a thread to access memory on the previous node, which degrades performance. Threads to be migrated must be selected effectively and efficiently since the related operations run in the critical path of the kernel scheduler. This study focuses on improving inter-node load balancing for multithreaded applications. We propose a thread-aware selection policy that considers the distribution of threads on nodes for each thread group while migrating one thread for inter-node load balancing. The thread is selected for which its thread group has the least exclusive thread distribution, and thread members are distributed more evenly on nodes. This has less influence on data mapping and thread mapping for the thread group. We further devise several enhancements to eliminate superfluous evaluations for multithreaded processes, so the selection procedure is more efficient. The experimental results for the commonly used PARSEC 3.0 benchmark suite show that the modified Linux kernel with the proposed selection policy increases performance by 10.7% compared with the unmodified Linux kernel.

Keywords: NUMA; Linux kernel; multithreaded; load balancing; remote memory access

Citation: Chiang, M.-L.; Su, W.-L. Thread-Aware Mechanism to Enhance Inter-Node Load Balancing for Multithreaded Applications on NUMA Systems. *Appl. Sci.* **2021**, *11*, 6486. https://doi.org/10.3390/app11146486

Academic Editor: Juan-Carlos Cano

Received: 8 June 2021
Accepted: 12 July 2021
Published: 14 July 2021

Publisher's Note: MDPI stays neutral with regard to jurisdictional claims in published maps and institutional affiliations.

1. Introduction

Multi-core systems allow parallel computing and have a higher throughput. To effectively utilize the performance of multi-cores, applications are coded as multithreaded. In Linux, the kernel maintains one runqueue for each core. When a process or thread is ready to run, it is put into the runqueue and waits to be run on the corresponding core. The Linux kernel [1] maintains a data structure, struct task_struct, which records attributes and runtime information for each schedulable entity. Each schedulable entity in the Linux kernel is called a task. When several tasks with different run times are run simultaneously, the load between cores can be imbalanced, so performance is decreased. The kernel scheduler's load balancing mechanism then migrates tasks from the overloaded core's runqueue to the runqueue of a core that is not so heavily loaded.

Non-Uniform Memory Access (NUMA) [2] systems divide system resources such as processors, caches, and RAM into several nodes. It takes longer for one core to access the memory on different NUMA nodes than on the local node. This costly memory access is called remote memory access. For a task running on a NUMA system, the memory pages allocated to it may be scattered on different nodes. When a task accesses memory pages on nodes other than those on which it runs, remote memory access occurs. For Linux-based NUMA systems, the load balancing mechanism can migrate tasks to another node, so costly remote memory access is necessary after the migration. The benefit of load balancing is reduced by the need for remote memory access after task migration. A prior study [3] showed that reducing remote memory access is critical in designing and implementing an operating system on NUMA systems.

In order to maintain a balanced load and reduce remote memory access, the kernel-level Memory-aware Load Balancing (kMLB) [4] mechanism was proposed to allow better inter-node load balancing for the Linux-based NUMA systems. The unmodified Linux kernel migrates the first movable task that can be run on the target core. The task may involve more remote memory access when it is migrated between nodes. The kMLB mechanism modifies the Linux kernel to track each task's number of memory pages on each node. This memory usage information is then used to select the task that is most suited to inter-node migration. Several task selection policies, such as Most Benefit (MB) [4] and Best Cost-Effectiveness [4], have been proposed and used different metrics. The selected tasks require less remote memory access after migration, and system performance successfully improves. Differently, Chen et al. [5] proposed a machine learning-based resource-aware load balancer in the Linux kernel to make migration decisions. The extra runtime overhead deducts the performance gain because scheduling operation is in the critical path of kernel operations.

A multithreaded process can create threads as needed during its execution. In the Linux kernel, these threads form one thread group and share memory space. On NUMA systems, threads of one thread group can be scheduled by the kernel to run on different nodes to balance the load, so the same memory pages can be accessed by threads that run on different nodes. Accessing one memory page involves local access for some threads and remote access for other threads. When a thread is migrated across nodes, remote memory access and cache misses increase as well, so it is difficult to determine the cost of memory access.

This study first analyzes multithreaded applications and their memory access in Linux and then proposes a new task selection policy, which is named Exclusivity (Excl) for multithreaded applications. This policy determines whether a task is suitable for inter-node migration using the exclusivity of the thread distribution on NUMA nodes in its thread group. The task for which the thread group is least exclusive is selected, which has a lesser effect on data mapping and thread mapping for its thread group.

Although selecting a suitable task for inter-node migration can reduce remote memory access, the selection procedure must evaluate all tasks in the runqueue, which involves a processing overhead. Since the selection is in the critical path for the kernel scheduler, the cost of more operations outweighs any benefit. Therefore, we further improve the procedure for selecting tasks by using thread group information to eliminate superfluous evaluations. Only a subset of movable tasks in the runqueue is evaluated, so the selection procedure is more efficient.

The contribution of this study is as follows. First, multithreaded applications and their memory access in Linux are analyzed. The analysis indicates that the existing memory-aware MB policy is still effective for multithreaded applications. However, it requires the kMLB mechanism to track per-task memory usage on per NUMA node. Thus, its implementation needs to modify many kernel operations and data structures that are affected. Second, the thread-aware Exclusivity policy is proposed, which is a relatively lightweight task selection policy since it does not need the kMLB mechanism. Instead, it uses the exclusivity of the thread distribution on nodes in the thread group to determine the target thread for inter-node migration. Third, several methods to enhance selecting tasks for inter-node migration for multithreaded applications are proposed.

Finally, the proposed Exclusivity policy and enhancement methods are practically implemented in the Linux kernel. Extensive experiments using the PARSEC 3.0 [6] benchmark suite run on the modified Linux kernel with various task selection policies. Compared with the unmodified Linux kernel, the results show that when the task selection procedure is enhanced, the Most Benefit Plus (MB$^+$) policy, which requires the kMLB mechanism, increases performance by 11.1%. The proposed Exclusivity policy increases performance by 10.7%. This policy is competitive and does not require the kMLB mechanism. Besides, it is more easily adapted to a newer Linux kernel.

The remainder of this paper is organized as follows. Section 2 introduces the technological background and related work. Section 3 presents the improvements for inter-node task migration for multithreaded applications and the new task selection policy. Section 4 details the experimental results, and Section 5 concludes.

2. Technological Background and Related Work

Though multi-core systems have high throughput, the performance increase depends on the placement of tasks and their data on nodes during runtime. Task placement affects contention between resources for the cache and cores, and data placement affects the memory access cost for a task. To utilize system resources more efficiently and increase performance, many studies [7–15] design specific placements of tasks and data to decrease resource contention, balance the loads in the cores, and reduce access to remote memory. The solutions focus on improving the performance of Symmetric Multi-processing (SMP) and Non-Uniform Memory Access (NUMA) [2] systems.

However, the rises of multithreaded applications cause finding specific task placement or data placement to increase performance more complicated. Since threads in one multithreaded application commonly share memory address space, and if they run on cores that do not share or share fewer cache resources, tasks are placed less efficiently because of more cache misses. In terms of data placement, there are different memory access latencies for cores access memory on different nodes. It is more challenging to determine where to allocate the required memory for the requesting task on NUMA systems.

This section first describes two types of multi-core systems and then reviews related work in Section 2.2. Section 2.3 introduces the kernel-based Memory-aware Load Balancing (kMLB) [4] mechanism and task selection policies proposed for improving inter-node migration.

2.1. Multi-Core Systems

A Symmetric Multi-processing (SMP) system is a multi-core system in which cores can share different levels of cache and the cost to access any location in the main memory is the same for all cores, as shown in Figure 1a. For a Non-Uniform Memory Access (NUMA) [1] system, as shown in Figure 1b, system resources such as cores, RAM, and memory controllers are divided into several nodes, and interconnection links connect these nodes. Cores can access the memory located on the same node with it, which is called local memory access. Accessing memory on other nodes via interconnection links is called remote memory access. This requires a longer time. The NUMA factor is the ratio between the remote memory access latency and the local memory access latency.

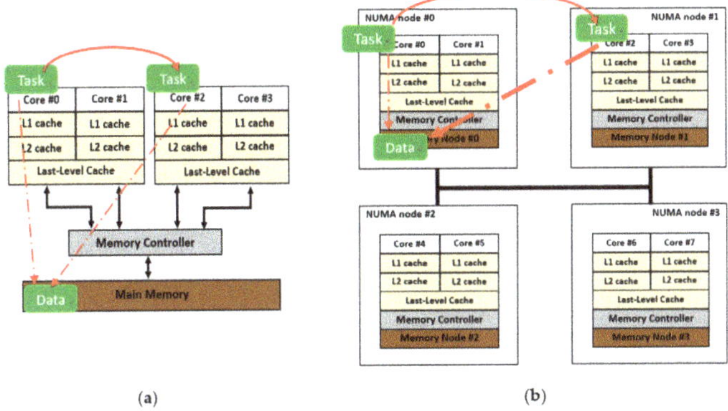

Figure 1. The load balancing mechanism affects SMP and NUMA systems differently. (a) SMP system; (b) NUMA system.

Though a NUMA system is more scalable than an SMP system and more memory accesses can occur simultaneously, operating system design must reduce costly remote memory access. Nevertheless, for load balancing, modern operating systems migrate tasks from an overloaded core's runqueue to the runqueue of a core with a lesser load. The load-balancing mechanism affects SMP and NUMA systems differently, as shown in Figure 1. Unlike an SMP system that features a uniform cost to access memory, migrating a task across nodes can involve costly remote memory access after migration. The benefit gained from load balancing thus is reduced by the need for remote memory access.

2.2. Related Work

In the Linux kernel, a memory page is allocated to a task when it first accesses the page. The page is allocated on the node where the requesting task is running. This is called a first-touch strategy [16]. Suppose a task is scheduled to run on another NUMA node during context switches or is migrated to another node for load balancing. In that case, the task requires remote memory access to access the page on the original node. Linux [17] offers NUMA-related system calls for NUMA-aware programs and provides commands and tools that constrain tasks to run on specific nodes. Several studies [11–14] proposed methods that use NUMA-related system calls to bind one task on specific nodes to decrease remote memory access. However, these performance improvements are reduced because a multithreaded application can create threads as needed during runtime, and the thread load is not deterministic. Binding tasks on specific nodes results in a load imbalance between nodes, and CPU utilization decreases accordingly.

Chen et al. [5] implemented a machine learning (ML)-based resource-aware load balancer in the kernel to make migration decisions. An ML model is implemented inside the kernel to monitor real-time resource usage in the system. This identifies potential hardware performance bottlenecks and then makes load balancing decisions. This ML model is trained offline in the user space and is used for online inference in the kernel to generate migration decisions. The results show no significant difference in the performance of the original kernel and the modified kernel when running benchmarks. Performance gains are negated largely because of the extra runtime overhead and the fact that scheduling operations are in the critical path of kernel operations.

Migrating memory pages to the NUMA node on which their requesting task is currently running reduces remote memory access. Mishra and Mehta [15] proposed an on-demand memory migration policy that migrates only the referenced pages to the current node where the requesting task is running. Terboven et al. [11] proposed a user-level implementation of a Next-touch approach in Linux. The *mprotect()* system call is used to change protection on a memory region, so the successive reads and writes incur segmentation faults. A signal handler that handles segmentation fault is implemented and invokes the *move_pages()* system call to migrate the accessed page to the node on which the task is currently running. Goglin and Furmento [13,14] presented two different implementations of a Next-touch approach in Linux. The user-space implementation also uses the *mprotect()* system call and a segmentation fault signal handler to migrate the accessed page. The kernel-level implementation uses the *madvise()* system call and modifies the kernel page fault handler to migrate the accessed page. The results show that the kernel-based implementation is more efficient than the user-space implementation. However, if the memory page is not accessed again after it is migrated, the cost of accessing the remote memory page may be less than the cost of migrating it.

In the Linux kernel, all threads of one multithreaded process share memory address space and use the same page table. If these threads run on different nodes, it is hard to determine whether pages should be migrated to the node where the requesting thread runs and to track the memory access pattern for an individual thread. Therefore, it is more challenging to perform thread mappings or data mappings to reduce remote memory access. To overcome these difficulties, Diener et al. [7] modified the kernel page fault handling routines to track the memory access patterns for any threads. The present flag of the page table entry is cleared so that whenever one memory page is accessed, a page fault occurs, even though the faulted memory page has already been in the memory. This identifies which thread on which node accesses this memory page and its access pattern. This mechanism is named kMAF [7] and uses the memory access patterns to determine which threads are more relevant and migrates them to the same node to allow better thread mappings. For data mappings, kMAF migrates one memory page to the node where the frequency of faults for this page is exclusive, so the page is mostly accessed from that node. This reduces remote memory access.

For methods using page fault to trigger migrating the faulted page to the same node as the faulting thread, the induced faults reduce performance. Existing studies also use page faults on the same page table for all threads for one multithreaded process to determine the memory access pattern for data mappings or thread mappings. Since several threads of a multithreaded process may fault one page, Gennaro et al. [8] indicated that this might result in an inaccurate estimation of the working-set of individual threads for one multithreaded process performance is decreased in terms of thread mappings. They then proposed a solution that uses the multi-view address space (MVAS). If MVAS is switched on while one multithreaded process runs, one individual page table is created for each of its threads. The memory access pattern for different threads can be separately tracked in different page tables until MVAS is switched off for the multithreaded process. MVAS does not incur extra page faults, so it can support those studies [11,13,14] that use page faults to perform page migrations to reduce remote memory access.

Lepers et al. [9] studied the placement of threads and data on NUMA nodes and the asymmetry of interconnecting links for nodes connected by links of different band-widths. A dynamic thread and memory placement algorithm was developed in Linux to minimize contention for asymmetric interconnect links and maximize bandwidth between communicating threads. Li et al. [10] also studied the effect of hardware asymmetry. The AMPS scheduler is implemented in the Linux kernel to support asymmetric multicore architectures for which cores in the same processor have different performances. The Linux kernel is modified to track the memory usage for each thread on each node and predicts the migration overhead for a thread. Threads are migrated to faster cores when they are under-utilized. However, if the predicted migration overhead is too high or the thread is in the memory allocation phase, this thread cannot be migrated across nodes.

2.3. The Kernel-Based Memory-Aware Load Balancing (kMLB) Mechanism

Inter-node task migration is necessary for an operating system to balance the load between NUMA nodes, and migrating different tasks for inter-node load balancing incurs a different amount of remote memory access. As shown in Figure 2, migrating the first task in the source runqueue, i.e., Task#3, incurs 3-page remote memory access but migrating Task#10 incurs 5-page remote memory access. However, the unmodified Linux kernel always selects the first task that can be migrated to run on the destination core. It allows rapid selection, but the first task may not incur the least remote memory access after the inter-node migration.

A previous study [4] shows that it is better to select the task that involves less remote memory access after migration. The kernel-based Memory-aware Load Balancing (kMLB) mechanism was proposed to select suitable tasks to migrate between nodes to allow better load balancing in the Linux kernel. The memory usage for each task on each node is tracked. Depending on a task's current running node, the physical pages that it occupies on each node are identified as local or remote. The load balancer then uses this information to determine the most suitable task for inter-node migration.

In the Linux kernel, the Resident Set Size (RSS) [1] for one process is the number of page frames occupied by this process. The original RSS only tracks the total number of page frames that are occupied by each process. The kMLB mechanism modifies the kernel operations to track the RSS counters on each node for each process, including dynamic memory allocation and releases, demand paging, swapping, system calls, and inter-node page migration.

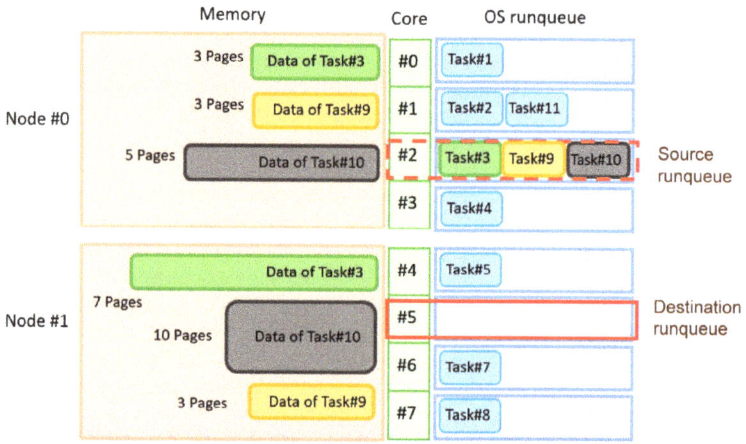

Figure 2. Migrating different tasks incurs a different degree of remote memory access after migration.

The following details task selection policies that are used with the kMLB mechanism. For the example in Figure 2, Table 1 shows the selected task for each policy according to the specific metric used.

Table 1. Selecting a task using different selection policies.

Policy	Tasks Waiting in the Source Core's Runqueue			Selected Task
	Task#3	Task#9	Task#10	
First-Fit	N/A	N/A	N/A	Task#3
TM	$3 + 7 = 10$	$3 + 3 = 6$	$5 + 10 = 15$	Task#9
MB	$7 - 3 = 4$	$3 - 3 = 0$	$10 - 5 = 5$	Task#10
BCE	$(7 - 3)/3 = 1.333$	$(3 - 3)/3 = 0$	$(10 - 5)/5 = 1$	Task#3

- Total Min (TM) Policy [3] selects the task in the source core's runqueue with the minimal total memory size. It is possible to have the least influence caused by task migration since the selected one occupies the least amount of memory for access after migration.
- Most Benefit (MB) Policy selects the task in the source core's runqueue that can reduce the maximum amount of remote memory access when migrated. This policy considers the memory that is occupied by each task on the source and destination nodes. The task with the maximum difference is selected. The following metric is used to select the target task with the maximum value, in which $RSS_p(i)$ is the RSS value for task p on NUMA node i and *dest_node* and *src_node* are the IDs for the destination and source NUMA nodes, respectively:

$$\text{MB}_p = \text{RSS}_p(dest\_node) - \text{RSS}_p(src\_node)$$

- Best Cost-Effectiveness (BCE) Policy selects the task in the source core's runqueue for which inter-node migration is the most cost-effective. Based on the MB policy, the BCE policy also considers the maximum cost of page migration, which is the amount of memory occupied by one task on the source node. The selected task is the one that can reduce the maximum remote memory access relative to the maximum migration cost when it is migrated. The following metric is used to select the target task with the maximum value:

$$\text{BCE}_p = (\text{RSS}_p(dest\_node) - \text{RSS}_p(src\_node))/\text{RSS}_p(src\_node)$$

3. Improved Inter-Node Load Balancing for Multithreaded Applications

The kMLB mechanism [4] tracks the number of physical pages per node occupied by each task. For each movable task in the overloaded core's runqueue, the modified OS scheduler uses this information to calculate the metric to determine the most suitable task for inter-node migration. However, the kernel scheduler must evaluate each task in the runqueue to identify a target task to be migrated. The source (i.e., the busiest) and the destination (i.e., the idlest) runqueues must be locked, so the target task must be identified efficiently because it runs in the critical path of the kernel scheduler. Besides, the threads of one multithreaded application share memory pages and may be distributed on different nodes. When threads running on different nodes access their shared memory pages on the remote node, cache misses and remote memory access slow access.

This study improves inter-node task migration for multithreaded applications in two respects. This section first introduces multithreaded applications and their memory access in Linux. Section 3.2 describes the improvements in selecting tasks for migration between nodes for multithreaded applications. Section 3.3 presents the proposed thread-aware task selection policy for inter-node migration for multithreaded applications.

3.1. Multithreaded Applications and Their Memory Access in Linux

During the execution of a multithreaded application, threads are created as needed. In Linux, these threads form one thread group. The first thread in a multithreaded process is the thread group leader, and other threads are the members of this thread group. Each thread is regarded as one schedulable entity in the Linux kernel, so it is one task. Threads of the same thread group share the same memory address space and page table.

When a task is created, the OS scheduler dispatches it to the core with the least load to maintain load balance within multi-core systems. On NUMA systems, threads of the same thread group may be distributed on different nodes, and their memory pages can also be allocated on several nodes, as shown in Figure 3. Therefore, memory pages are local memory pages for some threads and remote memory pages for others. Access to Data0 is local access for threads T0, T1, and T7 and remote access for threads T2, T3, T4, T5, and T6. The difference in the total memory access cost for a thread group when one of its threads is migrated across nodes must be determined.

As depicted in Figure 4, threads within one thread group can be scheduled to run on different nodes, and their memory spaces can be on different nodes. A thread group's total memory access cost sums up the local memory access costs and the remote memory access costs for threads in this thread group. For a NUMA system with n nodes for which the memory access latency from a remote node to the local node is constant, the number of threads within a specific thread group on the NUMA node i is denoted as N_i. The number of memory pages allocated to this thread group on the NUMA node i is denoted as R_i.

Regardless of the locality of memory references, the latency for all threads in the thread group to access the entire memory allocated to the thread group, which is the estimated total memory access cost, is shown in Equation (1). The local memory access cost is shown in Equation (2). Equation (3) shows the remote memory access cost, in which f is the NUMA factor that represents the ratio between the remote memory access latency and the local memory access latency.

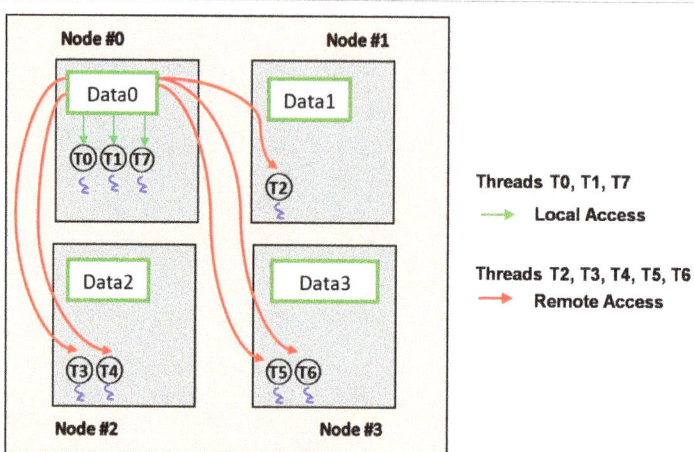

Figure 3. The memory pages and threads for a multithreaded process can be scattered on different nodes.

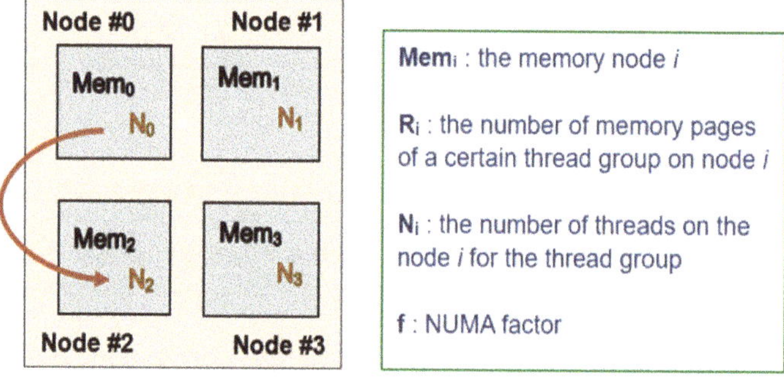

Figure 4. Migrating one thread of a specific thread group across nodes.

$$\text{Total memory access cost (TMA)} = \text{Local memory access cost (LMA)} + \text{Cache miss cost} + \text{Remote memory access cost (RMA)} \tag{1}$$

$$\text{Local memory access cost (LMA)} = \sum_{i=0}^{n-1} N_i * R_i \tag{2}$$

$$\text{Remote memory access cost (RMA)} = \left(\sum_{i=0}^{n-1} \sum_{\forall\ j \neq i} N_i * R_j\right) * f \tag{3}$$

Therefore, for the case in Figure 4, the estimated total memory access cost for this thread group is the value that is shown in Equation (4):

$$\begin{aligned}
\text{TMA} &= \sum_{i=0}^{3} N_i * R_i + \text{Cache miss cost} + \left(\sum_{i=0}^{3} \sum_{\forall\ j \neq i} N_i * R_j\right) * f \\
&= N_0 R_0 + N_1 R_1 + N_2 R_2 + N_3 R_3 + \text{Cache miss cost} + \\
&\quad (N_0 R_1 + N_0 R_2 + N_0 R_3 + N_1 R_0 + N_1 R_2 + N_1 R_3 + N_2 R_0 + N_2 R_1 + N_2 R_3 + N_3 R_0 + N_3 R_1 + N_3 R_2) * f
\end{aligned} \tag{4}$$

Because only the values of N_0 and N_2 change, and the others remain the same, the difference between the total memory access cost after a thread is migrated from node 0 to node 2 is simplified and calculated using Equation (5):

$$\text{Difference} = \text{TMA (after the migration of one thread from node 0 to node 2)} - \text{TMA (before the migration)}$$
$$= (R_2 - R_0) * (1 - f) \tag{5}$$

That is, if one thread is migrated, the difference between the total memory access cost after the migration is calculated using Equation (6), where R_D is the RSS value in the destination node, R_S is the RSS value in the source node, and f is the NUMA factor:

$$\text{Difference} = (R_D - R_S) * (1 - f) \tag{6}$$

Regarding inter-node task migration, the RSS values for a thread group on the source and the destination nodes have the most significant effect on the total memory access cost. Therefore, Most Benefit (MB) [4], which uses the same metric to select the most beneficial task is also appropriate for multithreaded applications.

3.2. Enhancements for Selecting Tasks for Inter-Node Migration for Multithreaded Applications

Selecting a suitable task for inter-node migration requires additional overhead because the selection procedure must evaluate all tasks in the runqueue in order. The evaluation cost increases as the number of tasks in the runqueue increases. For multithreaded applications, some evaluations are superfluous and can be eliminated because some tasks are less suitable than the candidate task. In this study, the thread group leaders are not migrated. The thread group leader's current node is determined when its thread group members are evaluated, and only one thread member per thread group is evaluated. Some tasks for the evaluation are eliminated if they match one of these aspects. Therefore, only the subset of movable tasks in the runqueue is evaluated, and the procedure for selecting tasks is more efficient. Figure 5 shows the flow for the improvements, and these methods are explained in the following subsections.

3.2.1. Eliminating the Thread Group Leader for Migration

The thread group leader is not selected because the Linux kernel uses the first-touch method [16] for physical memory allocation. A physical memory page is allocated the first time a thread accesses it and on the same node as the requesting thread. We observe the memory consumption for multithreaded applications, the first thread that touches the memory page is usually the thread group leader. The PARSEC 3.0 [6] benchmark suite contains 30 multithreaded applications, including 13 programs from PARSEC 2.0, 14 programs from the SPLASH benchmark suite, and three network programs. Table 2 shows these benchmark programs [4,6].

Each benchmark program's memory consumption was measured during its execution on the AMD server [18]. In Linux, the *free* command is used to obtain the current status for memory usage. A script that executes the free command every 0.1 s is used to record the memory usage footprint for the entire system. The increased memory usage during the benchmark program's execution is then attributed to the benchmark program. Each benchmark program is run with different threads, ranging from 1, 2, 4, 8, 16, and 32.

The results show that memory consumption is independent of the number of created threads for some benchmark programs. Other benchmark programs consume more memory as the number of created threads increases. Figure 6a shows the former situation for the benchmark program parsec.canneal. For this type of benchmark program, the memory pages of a multithreaded process are first touched by the first thread. For the benchmark program splash2x.fmm in Figure 6b, the individual threads first touch the memory pages of a multithreaded process. The results in Table 2 show that 22 of 30 applications allocate and initialize the allocated memory pages in the initializing thread (i.e., the thread group leader). The thread group leader for each multithreaded process is not migrated to reduce the scattering of memory pages on different nodes.

Table 2. The characteristics of benchmark programs in PARSEC 3.0.

Benchmark Suite	Benchmarks	Memory Allocated & Initialized
PARSEC 2.0	blackscholes, bodytrack, canneal, dedup, raytrace, streamcluster, swaptions	by thread group leader
	facesim, ferret, fluidanimate, freqmine, vips, x264	by individual threads
SPLASH-2x	barnes, cholesky, fft, lu_cb, lu_ncb, ocean_cp, ocean_ncp, radiosity, radix, raytrace, volrend, water_spatial	by thread group leader
	fmm, water_nsquared	by individual threads
Network	netdedup, netferret, netstreamcluster	by thread group leader

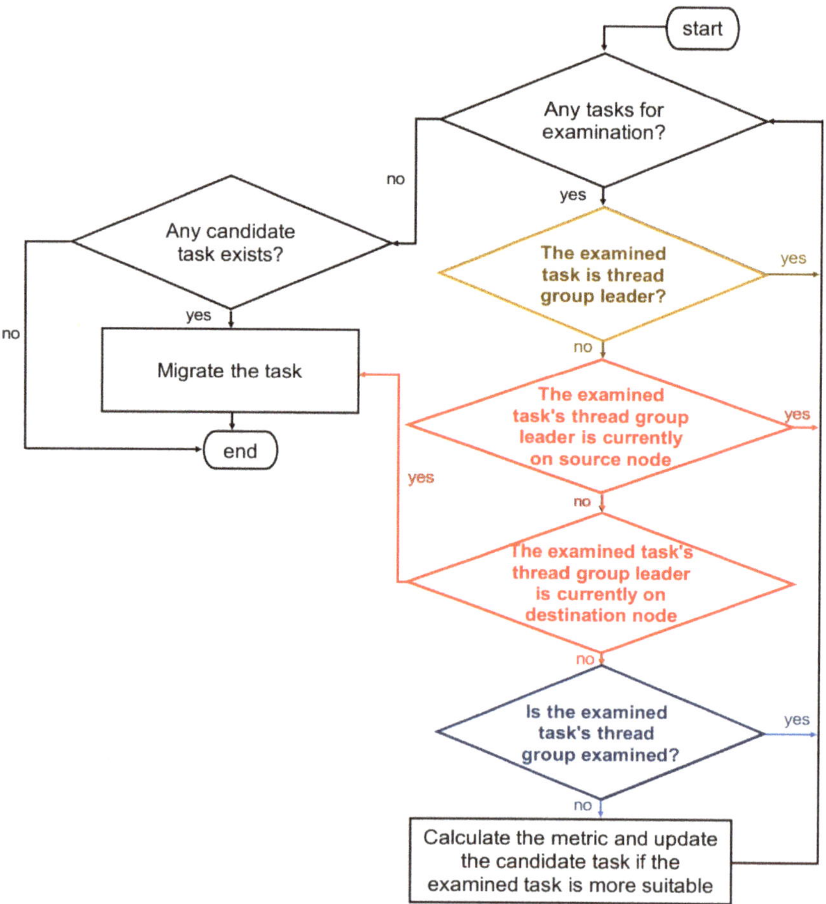

Figure 5. The modified flow for the efficient selection of tasks for inter-node migration.

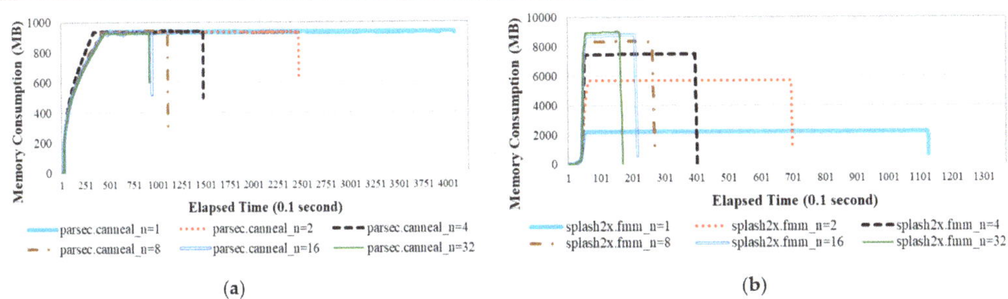

Figure 6. Memory consumption for benchmark programs running with different numbers of threads. (**a**) parsec.canneal; (**b**) splash2x.fmm.

3.2.2. Evaluating Only Those Tasks with Thread Group Leaders on Other Nodes

The observation presented in Section 3.2.1 shows that most of the memory pages allocated to a multithreaded process are on the node where the thread group leader is located. Therefore, ensuring that threads are executed on the same node as the thread group leader involves more local memory access.

For the proposed design, during the task evaluation for inter-node task migration, each task in the runqueue is classified into three types, according to where its thread group leader is currently located. Different decisions are made as follows. If the thread group leader is on the destination node, migrating this task to the destination node may involve more local memory access. Therefore, it is migrated directly instead of evaluating the remaining tasks in the runqueue. Suppose the thread group leader is on the source node. In that case, it is not selected for the migration because migrating it to the destination node may involve more remote memory access after migration. Therefore, only those tasks for which the thread group leaders are currently located on other nodes are evaluated.

Figure 7 illustrates several multithreaded processes running on a 4-node NUMA system. There are four thread groups and 19 tasks. The thread group leaders are denoted as "TGLR." Because Core#3 is idle, the OS scheduler performs the load balancing mechanism to migrate tasks from the overloaded runqueue (Core#1 on Node#0). The thread group leader for Task#3 is on the source node, so Task#3 is not selected for migration. Similarly, Task#11 is the target task and is migrated immediately because its thread group leader is on the destination node. If more tasks must be migrated to achieve load balancing, the remaining tasks (Task#19, Task#6, Task#16, and Task#9) are evaluated to determine which is best suited to inter-node migration.

3.2.3. Evaluating Only the First Thread Member in a Thread Group

In the source runqueue, only the first member in a thread group is evaluated in the selection procedure. This is because threads in the same thread group share the physical memory pages; their RSS counter values are the same. Figure 8 shows that the Linux kernel represents each thread with a *task_struct* structure, but all threads in the same thread group share the same memory management information (i.e., *mm_struct* structure) and page table. Their RSS counter values (i.e., *mm_rss_stat* structure) are also shared.

For the proposed design, while several tasks in the source runqueue are examined, only the first encountered thread member of a thread group is evaluated; other thread members in the same thread group are not evaluated. Therefore, the identical metric calculations to evaluate the threads of the same thread group in the runqueue are omitted. In the example of Figure 7, for Task#19, Task#6, Task#16, and Task#9, only Task#19 and Task#6 are evaluated.

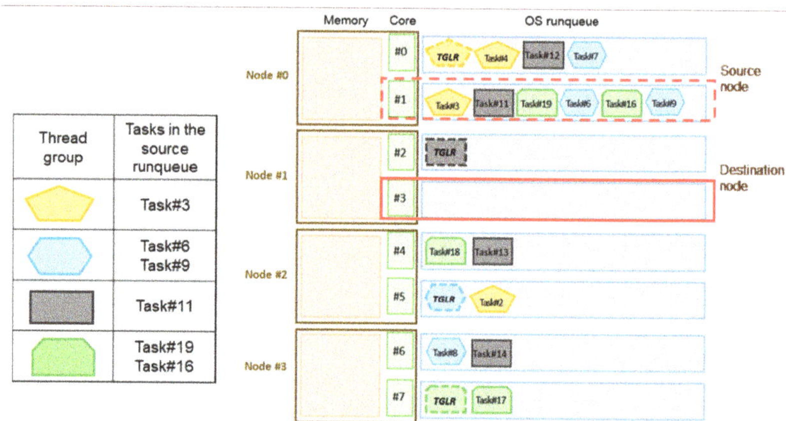

Figure 7. Several multithreaded processes running on a 4-node NUMA system.

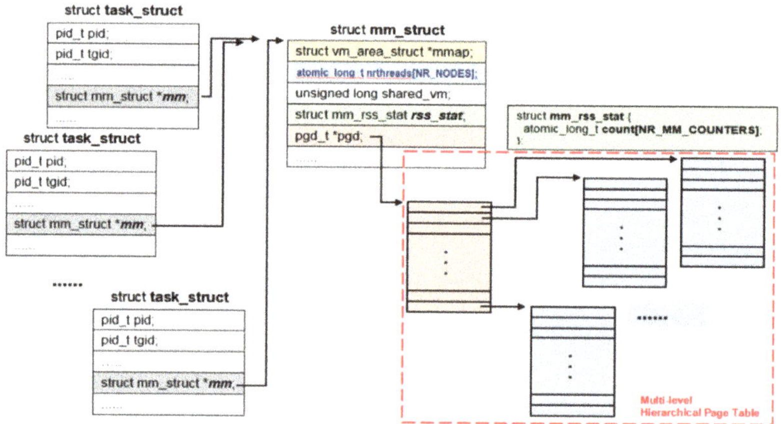

Figure 8. Tasks in a thread group share memory management information.

3.3. Task Selection Policy with Exclusivity for Multithreaded Applications

For existing policies, except for the First-Fit policy used in the unmodified Linux kernel, memory-aware policies, such as MB [4] and BCE [4] work with the kMLB mechanism. The Linux kernel must be modified to allow the kMLB mechanism to be used to determine the per-node memory pages per task. When the invoked functions or required data, structures are changed in the newer kernel, these memory-aware policies and the kMLB mechanism also require modification.

This study proposes a new thread-aware policy named Exclusivity (Excl) that does not require the kMLB mechanism. Instead, it considers the exclusivity of thread distribution on nodes for a thread group. The more evenly the threads of a thread group are distributed to nodes, the less beneficial it is for data mapping and thread mapping. Migrating a task across nodes also changes the thread distribution for the thread group. It is better to select a task for which its thread group's threads are distributed more evenly on nodes. The Excl policy selects the task for which the thread group is least exclusive in terms of thread distribution for inter-node migration.

Figure 9 illustrates an example in which 10 tasks belong to two thread groups. Most threads of one thread group are distributed on Node #0, but threads of the other group are evenly distributed on nodes. Migrating Task#10 is more beneficial.

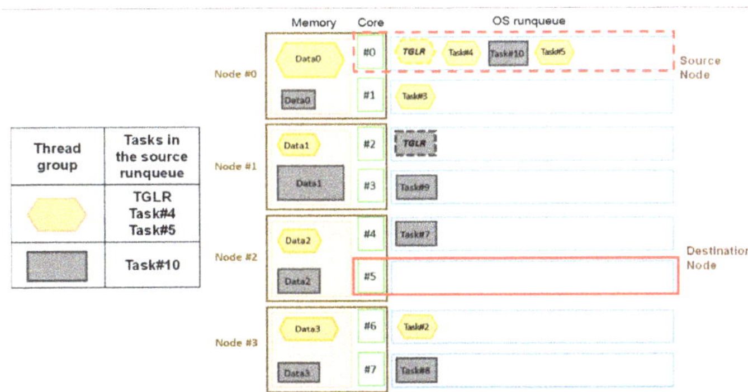

Figure 9. Two thread groups with different levels of exclusivity.

For the proposed policy using the thread distribution of its thread group, for each movable task p in the runqueue, Equation (7) is used to evaluate the exclusivity of this thread group. *thrd_nri* means the number of threads on NUMA node i, and n is the number of nodes. For tasks in the source core's runqueue, the task for which the thread group has the minimum value for exclusivity is selected. For tasks with equal exclusivity, the first one to be evaluated is the target task:

$$Excl_p = \max_{0 \leq i \leq n-1} thrd\_nr_i / \sum_{i=0}^{n-1} thrd\_nr_i \tag{7}$$

In Figure 9, threads in hexagon have the value of exclusivity 0.8, and threads in the rectangle have the value of exclusivity 0.4. In Figure 7, Task #11 is selected because it is the least exclusive. The evaluation is listed in Table 3.

Table 3. The evaluation of the Exclusivity policy. (**a**) The evaluation for tasks in Figure 9; (**b**) The evaluation for tasks in Figure 7.

(a)			
Thread Group	**Thread Group Members in Core#0 Runqueue**	$Excl_p$	**Selected Task**
hexagon	TGLR, Task#4, Task#5	$4/5 = 0.8$	Task#10
rectangle	Task#10	$2/5 = 0.4$	

(b)			
Thread Group	**Thread Group Members in Core#1 Runqueue**	$Excl_p$	**Selected Task**
pentagon	Task#3	$3/4 = 0.75$	Task#11
rectangle	Task#11	$2/5 = 0.4$	
shape	Task#19, Task#16	$2/5 = 0.4$	
hexagon	Task#6, Task#9	$3/5 = 0.6$	

However, exclusivity is not the sole criterion for consideration. There is an exceptional case for a multithreaded process for which most of the memory pages are allocated on some nodes, but most threads are on other nodes. In this situation, much remote memory access is necessary. As shown in Figure 10, the thread group in the hexagon and the thread group in the rectangle both have the same value of exclusivity 0.8. However, more remote memory access is necessary for the thread group in the hexagon. Suppose the task for which the thread group is least exclusive is selected. In that case, the highly exclusive thread group may still involve much remote memory access because the thread group leader is on a node different from those where most of the thread group members are located.

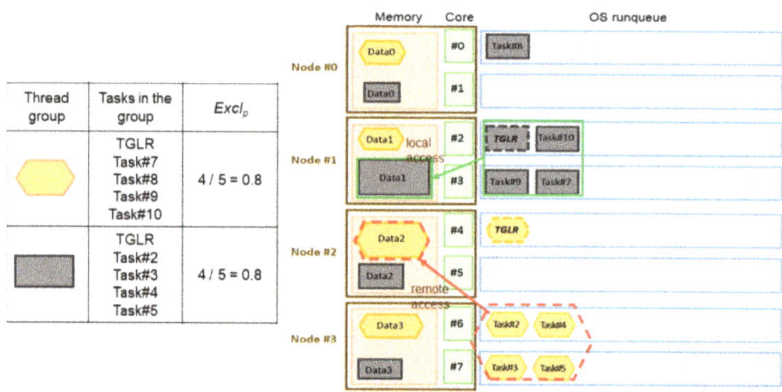

Thread group	Tasks in the group	$Excl_p$
	TGLR Task#7 Task#8 Task#9 Task#10	4 / 5 = 0.8
	TGLR Task#2 Task#3 Task#4 Task#5	4 / 5 = 0.8

Figure 10. A highly exclusive thread group may involve much remote memory access.

To ensure that threads and the data for a thread group are located on the same node, the proposed policy incorporates the consideration of a thread group leader in the task selection procedure. This makes members of a thread group remain on the same node as the thread group leader. Besides, as described in Section 3.2, most memory pages for a multithreaded process may be allocated by the thread group leader.

Each task in the runqueue is classified into three types according to the node where the thread group leader is currently located. If the thread group leader for a task is on the destination node, this task is not evaluated and is migrated immediately. A task for which its thread group leader is on the source node is not selected for migration. For tasks for which the thread group leader is on other nodes, Equation (7) is used to select the target task for inter-node migration.

4. Performance Evaluation

To measure the performance improvement due to the use of an enhanced inter-node load balancing procedure and the proposed policy for multithreaded applications, the benchmark suite PARSEC 3.0 [3] is used to test systems using different task selection policies and running benchmarks with various numbers of threads. The experiments record the elapsed running time for each test case, several performance counter events, and the elapsed running time for each benchmark program. The results are used to determine the reasons for an increase or decrease in performance.

Section 4.1 details the experimental environment, Section 4.2 describes the experimental design, and Section 4.3 presents the experimental results. Summary and discussions for experimental results are also provided.

4.1. Experimental Environment

The experiments are performed on the NUMA system, Supermicro A+ 4042G-72RF4 [18]. The software and hardware specifications are listed in Table 4. This is a 4-node NUMA system with a constant NUMA factor, and each node is installed with one AMD Opteron 6320 processor [19]. This processor has eight cores, so there is a total of 32 cores in the system. The kMLB [4] mechanism and task selection policies are implemented in the Linux kernel. The Linux *numactl* tool on the system shows that 1.6 times more access is required for remote memory than for local memory.

Table 4. Software and hardware specifications.

System	Supermicro A + 4042G-72RF4 [18] with 4 NUMA Nodes
Processor	4 AMD Opteron 6320 8-core Processor, 2.80 GHz
Motherboard	Supermicro H8QG7-LN4F
Memory controller	2 per node
Interconnect	6.4 GT/s AMD HyperTransport
Cache sizes per processor	L1 Data cache: 16 KB per core L1 Instruction cache: 64 KB per 2-core L2 cache: 2 MB per 2-core L3 cache: 8 MB among all cores per memory controller
Memory	DDR3 1600 16 GB per node (64 GB total)
Operating System	Ubuntu 13.10 Server Edition (Linux kernel 3.11.0-12-generic [20])

The PARSEC 3.0 [6] is a multithreaded benchmark suite that provides a convenient interface for building and running each benchmark program, as shown in Table 2 of Section 3.2.1. The multithreaded configuration "gcc-pthreads" is used to construct most benchmark programs, but the configuration "gcc-openmp" is used to construct the benchmark program parsec.freqmine. The input set "native" is used for performance analysis on real machines to run benchmark programs. The interface allows each benchmark program to be run with the specified number of threads.

4.2. Experimental Design

This study focuses on reducing remote memory access by improving inter-node load balancing. A sufficient number of benchmark programs must be run simultaneously on the experimental system, such that the kernel scheduler migrates tasks between nodes to balance the load when the nodes have an imbalanced load. 29 benchmark programs with a specific number of threads are run simultaneously for each test case. The numbers of threads range from 1, 2, 4, 8, 16, and 32. The elapsed running time for each test case and the elapsed running time for each benchmark program are measured using the performance counter statistics. For each test case, eight to ten runs are performed.

For each run, the system is rebooted to prevent buffer caching. During each run, the performance counter events in Table 5 are also recorded for each benchmark program. These are used to determine the cause of any change in performance: Instructions Per Cycle (IPC), Last Level Cache (LLC), and Miss Per Kilo Instructions (MPKI) to estimate runtime patterns. Other performance data is obtained from *numastat* command-line utility and *vmstat* in the *proc* file system.

Table 5. Collected performance counter events.

Event	Description
Elapsed Time	Used to evaluate the efficiency of running one benchmark program
CPU Cycles	Used to calculate the number of Instructions Per Cycle (IPC)
Instructions	Used to calculate IPC and Last Level Cache (LLC) Miss Per Kilo Instructions (MPKI)
LLC-load-misses	Used to calculate LLC MPKI. LLC-store-misses is not supported.
Page Faults	Number of page faults incurred by threads of one thread group
CPU Migrations	Number of task migrations for one thread group

This study also enhances existing task selection policies as described in Section 2.3 for multithreaded applications, so the experiments use different task selection policies, as shown in Table 6. The First-Fit policy is used in the unmodified Linux kernel. As presented in Section 3.1, the MB [4] policy considers the memory occupied by one task on the destination and source nodes to select a target task for inter-node migration. This policy is suited to multithreaded applications. However, additional overhead is incurred because the policy relies on the information provided by the kMLB mechanism [4].

Table 6. The experimental system running different task selection policies.

Experimental System	Task Selection Policy	Need kMLB Mechanism	Features
(A)	Default	No	The unmodified Linux kernel with the first-fit policy.
(B)	MB	Yes	The modified Linux kernel with kMLB mechanism and MB policy.
(C)	MB$^+$	Yes	Based on the MB policy. Enhanced task selection procedure.
(D)	Excl$_{base}$	No	The modified Linux kernel that selects the task for which the thread group has the minimum value of exclusivity.
(E)	Excl	No	Based on the Excl$_{base}$ policy. Enhanced task selection procedure.

To allow faster selection of tasks, we also enhance the MB policy to be MB$^+$. Each task in the runqueue is evaluated only when its thread group leader is currently not located on the destination or the source nodes. Therefore, only the subset of tasks in the runqueue must be evaluated. Besides, a thread group leader is not migrated across nodes, and a task is migrated directly if its thread group leader is on the destination node.

4.3. Experimental Results

4.3.1. Performance Comparison for Varying Numbers of Threads

The benchmark programs were run with varying numbers of threads, ranging from 1, 2, 4, 8, 16, and 32. The average elapsed time, the standard deviation, and the ratio of these times to an unmodified Linux kernel are calculated for different task selection policies and the specific number of threads. Figure 11 shows the results.

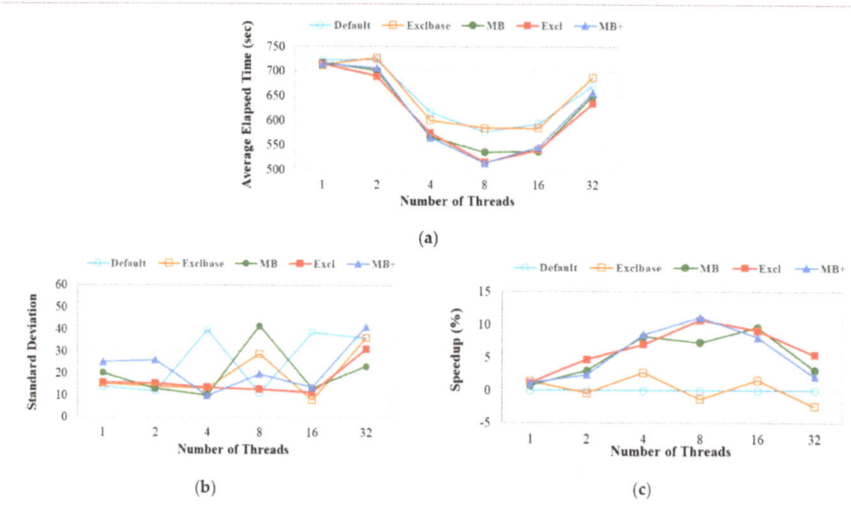

Figure 11. Experimental results for different numbers of threads. (**a**) Average elapsed time of each test case; (**b**) Standard deviation of each test case; (**c**) The performance increase over the unmodified Linux kernel.

Figure 11a shows that all test cases behave similarly in terms of average elapsed time for different numbers of threads. As the number of threads increases, multithreaded benchmarks run faster since a multi-core system's parallel computing capability increases performance as well. Besides, since more threads wait in the runqueue, an effective task selection policy can select a more suitable one among them for migration. The proposed task selection policies allow more efficient inter-node task migration for the load balancing mechanism on the experimental NUMA system. These perform better than the First-Fit policy that is used in the unmodified Linux kernel.

However, the parallel computing capability of a multi-core system is limited, the elapsed time measured depends on the number of cores in the target system. Because the experimental system has 32 cores, if there are too many threads for a multithreaded benchmark, contention for cores and memory slows down the performance. On the contrary, if the number of threads in a multithreaded benchmark is much smaller, multi-core is not fully utilized. Besides, few or no threads wait in runqueue, then task selection policies are not triggered or used effectively. Therefore, the increase in performance is not so great when the numbers of threads are 1, 2, and 32, as shown in Figure 11c.

4.3.2. The Effect of Enhancing the Task Selection Procedure

The procedure for selecting tasks runs in the critical path of the kernel scheduler, so the target task for the migration must be identified as efficiently as possible. Therefore, this study selects the target task more quickly to allow more efficient inter-node task migration for multithreaded applications. Only the task in the runqueue for which the thread group leader is currently on nodes rather than destination and source nodes is evaluated. No thread group leader is migrated across nodes.

Figure 12 shows the different improvement ratios. The $Excl_{base}$ policy achieves a more significant improvement than the MB policy. This enhancement has a different effect on each because the MB policy evaluates each task in the runqueue to select the most beneficial task for migration. For the MB^+ policy, though this enhancement allows faster task selection by only evaluating the subset of tasks in the runqueue, the selected target task may not result in a greater benefit than the most beneficial task. If there are a few tasks in the runqueue, the saved evaluation costs can be negated by selecting a target task that is not the most suitable.

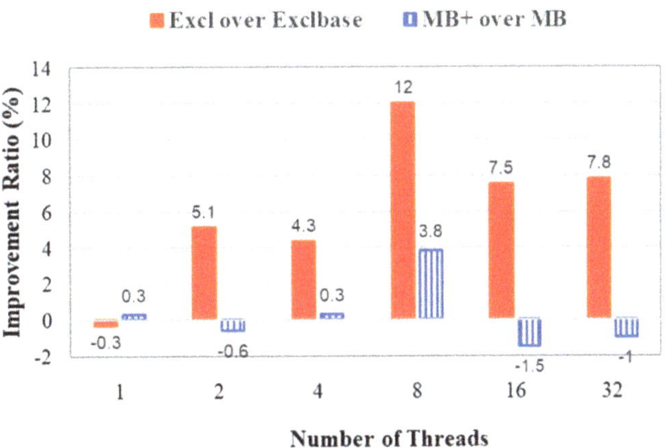

Figure 12. Improvement ratio for enhancing the task selection procedure.

The $Excl_{base}$ policy selects the task for which the thread group is least exclusive regarding the thread distribution on nodes. The $Excl_{base}$ policy does not consider the examined task's thread group leader, so the selected task can be migrated to the node different from where its thread group leader is currently located. Moving tasks away from its thread group leader can involve more remote memory access after migration. In contrast, the Excl policy incorporates the proposed enhancements for task selection procedure into the Excl$_{base}$ policy. The Excl policy evaluates the task for which the thread group leader is on other nodes and migrates the task to the node the thread group leader is on. Most memory pages are first touched by the thread group leaders, as presented in Section 3.2, so the Excl policy successfully enhances the Excl$_{base}$ policy and improves the performance.

4.3.3. Performance Counter Statistics

We analyze the causes of performance improvements using the measurement data obtained from the performance counter. Since 29 benchmark programs were run with varying numbers of threads, ranging from 1, 2, 4, 8, 16, and 32. They were run on the NUMA server with the Linux kernels using different task selection policies. Lots of figures are obtained from experimental results. From the measurement of the standard deviation of each test case, shown in Figure 11b, the standard deviations for test cases using 4 or 16 threads are relatively small for various task selection policies. Since there are 32 cores in the system, a sufficient number of threads must be run simultaneously on the system. The kernel scheduler then migrates threads between nodes to balance the load when the nodes' loads are imbalanced. As the number of threads increases, more threads wait in the runqueue, and an effective task selection policy can select a more suitable one for migration. Therefore, we observe the runtime patterns for these test cases that use 16 threads. Figure 13 shows the result.

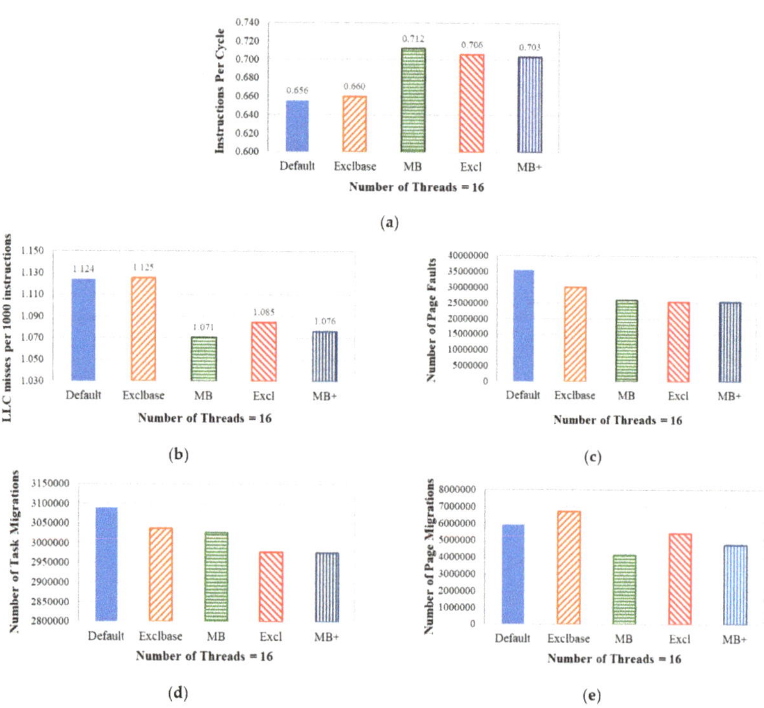

Figure 13. Experimental results for systems using different policies. (**a**) Instructions per CPU cycle; (**b**) LLC misses per 1000 instructions; (**c**) The number of page faults; (**d**) The number of tasks that are migrated; (**e**) The number of pages that are migrated.

Figure 13a,b shows the Instructions Per Cycle (IPC) and the Last Level Cache (LLC) misses per 1000 instructions (LLC MPKI) for systems that use different task selection policies. IPC is used as a reference to evaluate the CPU utilization rate. Cores fetch instructions or data from caches for execution. If the LLC misses, the required instructions or data are then accessed from the main memory. Hence, the LLC MPKI is used as a reference to evaluate the frequency of memory access. The results show that systems using the proposed task selection policies outperform the unmodified Linux kernel, which uses the default First-Fit policy.

We also record the number of page faults, task migrations, and page migrations for each test case, and the results are respectively shown in Figure 13c,d. During the execution of tasks, page faults, task migrations, and page migrations increase the runtime overhead. Systems that use the proposed task selection policies all incur fewer of these operations and have a lower overall runtime overhead than the unmodified Linux kernel, so they perform better.

4.3.4. Performance Results for Various Benchmarks

The performance improvement is measured for each benchmark program. Figure 14 shows the results for each benchmark program running on systems that use different task selection policies for the test cases with 16 threads. We observe that the benchmark programs for which the running time is longer (e.g., parsec.facesim and splash2x.raytrace) obtain more performance gains on systems that use the proposed task selection policies. The benchmark programs (e.g., splash2x.cholesky and parsec.vips) show no performance improvement because their running time is too short. Tasks requiring a longer elapsed time have more chances to be migrated and thus gain more performance improvement.

4.3.5. Summary and Discussion

As the analysis presented in Section 3.1, MB is still effective for multithreaded applications. The experimental results demonstrate that MB and MB$^+$ achieve good performance. The experimental results also demonstrate that the Exclusivity policy is competitive. Although its 10.7% performance improvement over the unmodified Linux kernel is still slightly less than the 11.1% performance improvement over the same for MB$^+$ with the kMLB mechanism.

As presented in the study [4], the MB policy works with the kMLB mechanism, and the implementation includes two major works. First, the kernel's memory management routines and exception handling routines are modified to obtain per-task memory usage on each node. Second, the kernel's inter-node load balancing procedure is modified to incorporate the task selection policy. Therefore, to support the kMLB mechanism in the Linux kernel, all kernel operations that update the values of RSS counters and related data structures are modified to track the RSS counters on each node for each process. In detail, seven types of operations are affected and modified, regarding dynamic memory allocation and releases, demand paging, copy-on-write mechanism, swapping, related system calls, and inter-node page migration. However, the latest version of the Linux kernel [20] still does not support the separate counting of memory usage on each node for each process.

In contrast, the Exclusivity policy does not require the kMLB mechanism. Instead, only the kernel's inter-node load balancing procedure is modified to incorporate the task selection policy. Thus, adapting the Exclusivity policy to a newer Linux kernel is less complicated than implementing memory-aware policies with the kMLB mechanism.

On the other hand, the proposed thread-aware mechanism aims to enhance inter-node load balancing for multithreaded applications on NUMA systems. Therefore, it has some limitations. A sufficient number of threads must be run simultaneously on the system, such that the kernel scheduler then migrates threads between nodes to balance the load when the nodes' loads are imbalanced. As the number of threads increases, more threads are likely to wait in the runqueue, and an effective task selection policy can select a more suitable one among them for migration. In contrast, the default Linux kernel always migrates the first task in the runqueue. Therefore, as shown in the experimental results, its performance is not stable and not good since the first task may not be a good one for migration.

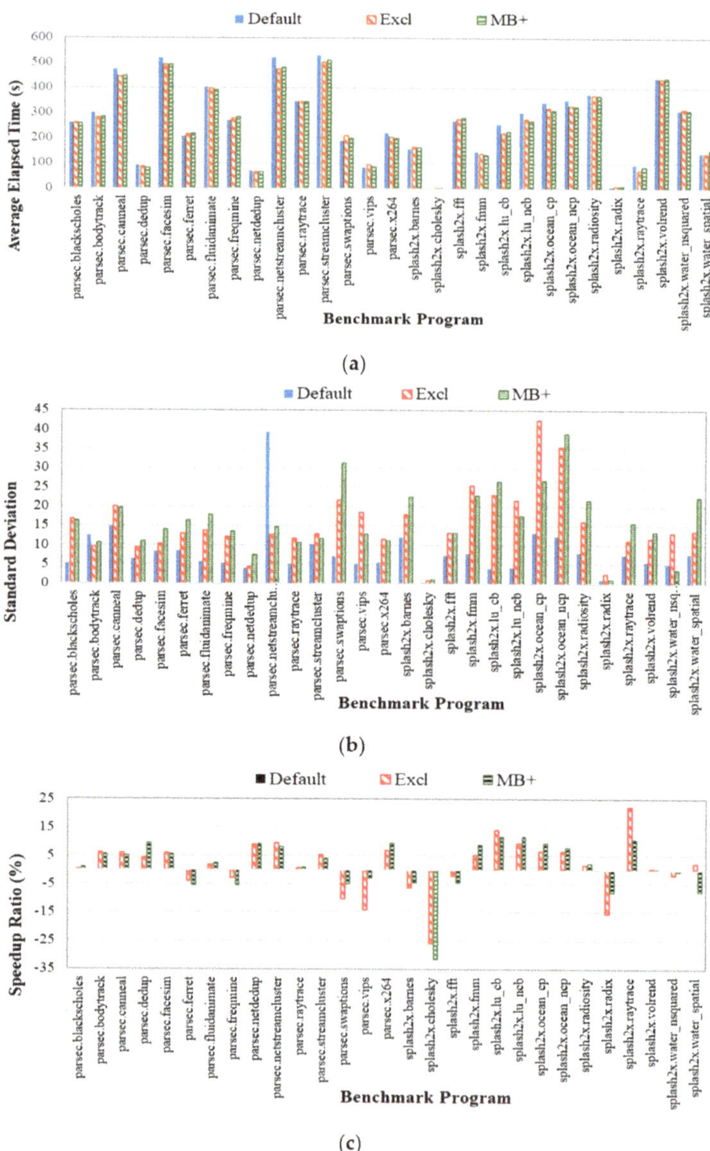

Figure 14. Experimental results for systems running each benchmark program. (**a**) The average elapsed time; (**b**) The standard deviation in the running time; (**c**) The speedup ratio.

However, the parallel computing capability of a multi-core system is limited and depends on the number of cores in the target system. If there are too many threads running, contention for cores and memory resource slows down the performance. On the contrary, if there are too few or no threads wait in runqueue, task selection policies are not triggered or used effectively. The experimental results show that the increase in performance is not significant when the numbers of threads are few or too many.

Regarding the power conservation issue, most modern CPUs support many frequencies. The higher the clock frequency, the more energy is consumed over a unit of time. The longer a thread is running, the more energy is consumed as well.

The Linux kernel supports CPU performance scaling through the CPUFreq (CPU Frequency scaling) subsystem [21]. In our experiments, the Linux kernel uses the default setting of CPU frequency governor "ondemand" [21], which sets the CPU frequency depending on the current usage for all test cases. Since many benchmark programs with many threads are running simultaneously, the system load is high. Therefore, CPUs are set by the Linux kernel to run at the highest frequency during experiments. The experimental results show that benchmark programs run 10.7% faster on the modified Linux kernel with the proposed task selection policy than on the default Linux kernel. Under the same CPU frequency, the shorter the benchmark programs run, the less energy is consumed. The degree of energy-saving needs further experiments.

5. Conclusions

Multi-core systems feature a high throughput, but load imbalance can degrade performance. For NUMA multi-core systems, there is non-uniform memory access, so migrating tasks across nodes to achieve load balancing has a different memory access cost. Therefore, the tasks to be migrated must be selected effectively and efficiently, especially the related operations run in the critical path of the kernel scheduler.

On Linux-based NUMA systems, threads of a multithreaded application share the memory address space and can be scheduled to run on different nodes. Memory pages allocated to them can also be on different nodes. Several studies present specific mechanisms to adjust threads and memory pages on nodes to reduce remote memory access and achieve load balancing. However, strategies using page fault handling to migrate threads or memory pages induce certain overhead. Besides, the kernel scheduler's inter-node task migration can mess up the arrangement. Differently, the kernel-level kMLB [4] mechanism enhances inter-node load balancing for NUMA systems, which tracks the number of memory pages on each node occupied by each task. This memory usage information is then used by memory-aware task selection policies [4] to select the most suitable task for inter-node migration. Despite the required overhead, the kMLB mechanism with task selection policies increases the performance of NUMA systems.

This research studies the memory access for multithreaded processes and proposes a thread-aware kernel mechanism to enhance inter-node load balancing for multithreaded applications on NUMA systems. The proposed Exclusivity policy migrates the task for which its thread group is least exclusive in the thread distribution. A thread group for which tasks are distributed more evenly on different nodes has less impact after task migration. The enhanced task selection procedure does not select the thread group leader for migration to prevent memory pages for one multithreaded process from being scattered on multiple nodes. Besides, only those tasks for which their thread group leaders are on other nodes are evaluated. The proposed policy allows threads of the same thread group to remain on the same node, so performance increases.

This study shows that the Most Benefit (MB) policy is still effective for multithreaded applications, and the proposed Exclusivity policy is competitive with the MB policy. Compared with unmodified Linux, the system that uses the MB$^+$ policy with the kMLB mechanism increases performance by 11.1%. The system that uses the Exclusivity policy, which does not require the kMLB, increases performance by 10.7%. In comparison, it is less complicated to adapt the Exclusivity policy to a newer Linux kernel than to use memory-aware policies with the kMLB mechanism. Moreover, under the same CPU frequency, the shorter the programs run, the less energy is consumed. We plan to adapt our work to a newer Linux kernel and perform experiments on more NUMA systems and energy saving in the future.

Author Contributions: Methodology, M.-L.C. and W.-L.S.; software, W.-L.S.; writing—original draft preparation, M.-L.C. and W.-L.S.; writing—review and editing, M.-L.C.; project administration, M.-L.C.; funding acquisition, M.-L.C. All authors have read and agreed to the published version of the manuscript.

Funding: This research was funded in part by Ministry of Science and Technology, Taiwan, grant number 106-2221-E-260-001, 107-2221-E-260-005, 108-2221-E-260-005, and 109-2221-E-260-011.

Acknowledgments: We would like to thank Y. L. Sam for his assistance in performing extensive experiments. Special thanks to the anonymous reviewers for their valuable comments in improving this research.

Conflicts of Interest: The authors declare no conflict of interest.

References

1. Bovet, D.P.; Cesati, M. *Understanding the Linux Kernel*, 3rd ed.; O'Reilly Media Inc.: Sebastopol, CA, USA, 2005; ISBN 0596005652.
2. Lameter, C. An Overview of Non-Uniform Memory Access. *Commun. ACM* **2013**, *56*, 59–65. [CrossRef]
3. Chiang, M.L.; Tu, S.W.; Su, W.L.; Lin, C.W. Enhancing Inter-Node Process Migration for Load Balancing on Linux-based NUMA Multicore Systems. In Proceedings of the 10th IEEE International Workshop on Computer Forensics in Software Engineering, Tokyo, Japan, 23–27 July 2018.
4. Chiang, M.L.; Su, W.L.; Tu, S.W.; Lin, Z.W. Memory-Aware Kernel Mechanism and Policies for Improving Inter-Node Load Balancing on NUMA Systems. In *Software: Practice and Experience*; John Wiley & Sons Ltd.: Hoboken, NJ, USA, 2019; Volume 49, pp. 1485–1508.
5. Chen, J.; Banerjee, S.S.; Kalbarczyk, Z.T.; Iyer, R.K. Machine Learning for Load Balancing in the Linux Kernel. In Proceedings of the 11th ACM SIGOPS Asia-Pacific Workshop on Systems, Tsukuba, Japan, 24–25 August 2020.
6. PARSEC Benchmark Suite. Available online: http://parsec.cs.princeton.edu/ (accessed on 1 June 2021).
7. Diener, M.; Cruz, E.H.M.; Alves, M.A.Z.; Navaux, P.O.A.; Busse, A.; Heiss, H.U. Kernel-Based Thread and Data Mapping for Improved Memory Affinity. In *IEEE Transactions on Parallel and Distributed Systems*; IEEE: Piscataway, NJ, USA, 2016; Volume 27, pp. 1–14.
8. Gennaro, I.D.; Pellegrini, A.; Quaglia, F. OS-based NUMA Optimization: Tackling the Case of Truly Multithread Applications with Non-partitioned Virtual Page Accesses. In Proceedings of the IEEE/ACM International Symposium on Cluster, Cloud, and Grid Computing, Cartagena, Colombia, 16–19 May 2016; pp. 291–300.
9. Lepers, B.; Quéma, V.; Fedorova, A. Thread and Memory Placement on NUMA Systems: Asymmetry Matters. In Proceedings of the 2015 USENIX Annual Technical Conference, Santa Clara, CA, USA, 8–10 July 2015.
10. Li, T.; Baumberger, D.; Koufaty, D.A.; Hahn, S. Efficient Operating System Scheduling for Performance-asymmetric Multi-core Architectures. In Proceedings of the 2007 ACM/IEEE Conference on Supercomputing, Reno, NV, USA, 10–16 November 2007.
11. Terboven, C.; Mey, D.A.A.; Schmidl, D.; Jin, H.; Reichstein, T. Data and Thread Affinity in OpenMP Programs. In Proceedings of the 2008 Workshop on Memory Access on Future Processors, Ischia, Italy, 5 May 2008; pp. 377–384.
12. Unat, D.; Dubey, A.; Hoefler, T.; Shalf, J.; Abraham, M.; Bianco, M.; Chamberlain, B.L.; Cledat, R.; Edwards, H.C.; Fuerlinger, K.; et al. Trends in Data Locality Abstractions for HPC Systems. In *IEEE Transactions on Parallel and Distributed Systems*; IEEE: Piscataway, NJ, USA, 2017; Volume 28, pp. 3007–3020.
13. Goglin, B.; Furmento, N. Memory Migration on Next-touch. In Proceedings of the Linux Symposium, Montreal, QC, Canada, 13–17 July 2009.
14. Goglin, B.; Furmento, N. Enabling High-Performance Memory Migration for Multithreaded Applications on Linux. In Proceedings of the IEEE International Symposium on Parallel & Distributed Processing, Rome, Italy, 23–29 May 2009.
15. Mishra, V.K.; Mehta, D.A. Performance Enhancement of NUMA Multiprocessor Systems with On-Demand Memory Migration. In Proceedings of the 2013 IEEE 3rd International Advance Computing Conference, Ghaziabad, India, 22–23 February 2013; pp. 40–43.
16. Marchetti, M.; Kontothanassis, L.; Bianchini, R.; Scott, M. Using Simple Page Placement Policies to Reduce the Cost of Cache Fills in Coherent Shared-memory Systems. In Proceedings of the 9th International Parallel Processing Symposium, Santa Barbara, CA, USA, 25–28 April 1995.
17. Schermerhorn, L.T. Automatic Page Migration for Linux. In Proceedings of the Linux Symposium, Sydney, Australia, 15–20 January 2007.
18. Supermicro AS 4042G-72RF4. Available online: https://www.supermicro.com/Aplus/system/Tower/4042/AS-4042G-72RF4.cfm (accessed on 1 June 2021).
19. AMD Opteron 6320 Processor. Available online: https://www.amd.com/en/products/cpu/6320 (accessed on 1 June 2021).
20. The Linux Kernel Archives. Available online: https://www.kernel.org/pub/linux/kernel/v3.0/ (accessed on 1 June 2021).
21. Linux CPUFreq Governors—Information for Users and Developers. Available online: https://www.kernel.org/doc/Documentation/cpu-freq/governors.txt (accessed on 4 July 2021).

Article

A Multi-Strategy Improved Sparrow Search Algorithm for Solving the Node Localization Problem in Heterogeneous Wireless Sensor Networks

Hang Zhang [1], Jing Yang [1,2,*], Tao Qin [1], Yuancheng Fan [3], Zetao Li [1] and Wei Wei [4,*]

[1] Electrical Engineering College, Guizhou University, Guiyang 550025, China; gs.zhanghang19@gzu.edu.cn (H.Z.); tqin@gzu.edu.cn (T.Q.); gzulzt@163.com (Z.L.)
[2] Key Laboratory of Advanced Manufacturing Technology of Ministry of Education, Guizhou University, Guiyang 550025, China
[3] China Power Construction Group, Guizhou Engineering Co., Ltd., Guiyang 550002, China; fanyc-gzgc@powerchina.cn
[4] China Power Construction Group, Guizhou Electric Power Design and Research Institute Co., Ltd., Guiyang 550002, China
* Correspondence: jyang7@gzu.edu.cn (J.Y.); weiwei-gzy@powerchina.cn (W.W.)

Abstract: Aiming at the problems of slow convergence and low accuracy of the traditional sparrow search algorithm (SSA), a multi-strategy improved sparrow search algorithm (ISSA) was proposed. Firstly, the golden sine algorithm was introduced in the location update of producers to improve the global optimization capability of SSA. Secondly, the idea of individual optimality in the particle swarm algorithm was introduced into the position update of investigators to improve the convergence speed. At the same time, a Gaussian disturbance was introduced to the global optimal position to prevent the algorithm from falling into the local optimum. Then, the performance of the ISSA was evaluated on 23 benchmark functions, and the results indicate that the improved algorithm has better global optimization ability and faster convergence. Finally, ISSA was used for the node localization of HWSNs, and the experimental results show that the localization algorithm with ISSA has a smaller average localization error than that of the localization algorithm with other meta-heuristic algorithms.

Keywords: sparrow search algorithm; gold sine algorithm; Gaussian disturbance; heterogeneous wireless sensor networks; node localization

Citation: Zhang, H.; Yang, J.; Qin, T.; Fan, Y.; Li, Z.; Wei, W. A Multi-Strategy Improved Sparrow Search Algorithm for Solving the Node Localization Problem in Heterogeneous Wireless Sensor Networks. *Appl. Sci.* **2022**, *12*, 5080. https://doi.org/10.3390/app12105080

Academic Editor: Peng-Yeng Yin

Received: 12 April 2022
Accepted: 16 May 2022
Published: 18 May 2022

Publisher's Note: MDPI stays neutral with regard to jurisdictional claims in published maps and institutional affiliations.

1. Introduction

Wireless sensor networks (WSNs) consist of a large number of miniature sensor nodes with low energy consumption, low price, and reliable performance [1]. They are often used in environmental monitoring, geological disaster warning, military reconnaissance, and other fields [2]. In these fields, location information is crucial, as data without geographic coordinate information are worthless [3]. Usually, a WSNs consists of hundreds or even thousands of sensor nodes, and designers cannot guarantee that all sensor nodes are of the same model. Further, the signal transmission power of the sensors generated by different sensor manufacturers will be different, which leads to the heterogeneity of the communication radius of sensor nodes. There may be various reasons for the formation of heterogeneous wireless sensor networks (HWSNs), but for localization techniques, the most important concern is the heterogeneity of the node communication radius. The localization problem of heterogeneous wireless sensor networks is similar to that of homogeneous wireless sensor networks in that the coordinates of unknown nodes are calculated by a specific localization algorithm using anchor nodes containing geographic coordinates within the network. However, unlike homogeneous wireless sensor networks, the heterogeneity of the node communication radius leads to a further increase in the localization error. Moreover, there are relatively few studies on conducting node localization of HWSNs.

Currently, researchers have proposed many localization algorithms. With the exception of the centroid localization algorithm, most localization algorithms can be divided into two stages: distance estimation and coordinate calculation. In the distance estimation phase, researchers can use the signal propagation time, the attenuation value of signal strength from the sending node to the receiving node, or the average hop distance and hop count between sensor nodes to calculate the distance between the unknown node and each anchor node [4]. In the coordinate calculation phase, the most commonly used method is the least-squares method (LS).

In recent years, meta-heuristic algorithms, which are known for their simplicity, flexibility, and spatial search capability, have provided a new idea for node coordinate calculation in WSNs. Liu et al. [5] replaced LS with a modified particle swarm algorithm (M-PSO) for the coordinate calculation of unknown nodes; when the error in the distance estimation stage was less than 10%, the error using M-PSO was 12.613 m less than that of the localization algorithm using LS. Chai et al. [6] proposed a parallel whale optimization algorithm to replace LS in the DV-Hop algorithm, and the obtained localization error was reduced by 8.4% compared with the DV-Hop algorithm. Cui et al. [7] improved the DV-Hop algorithm, made the discrete hop values continuous, and used the differential evolution algorithm (DE) to replace LS in the coordinate calculation stage. The positioning error of the algorithm was reduced by 70% compared with the DV-Hop algorithm. Although higher localization accuracy can be obtained using the meta-heuristic algorithm compared with LS, most of these studies focus on homogeneous wireless sensor networks, and few researchers have focused on HWSNs.

So far, not many results have been achieved on node localization of HWSNs. Assaf et al. [8] proposed a new expected hop progress (EHP) localization algorithm applicable to nodes with different transmission capabilities. The distance estimated by this algorithm in the distance estimation phase is closer to the real distance. However, the requirements for the potential forwarding area of the successor node are relatively high, and the communication radius of the sensor node cannot be too large. Wu et al. [9] optimized EHP and used elliptic distance to correct the distance calculated by the forward hop progress method and then used LS to find the coordinates of the unknown node. However, the node density required by the algorithm is too large. What is more, there is still room for improving the positioning accuracy of the nodes. Bhat et al. [10] proposed a minimum area localization algorithm for HWSNs combined with the Harris Hawk optimization algorithm, but the algorithm lacks a comparison with the use of other metaheuristic algorithms, and it is difficult to highlight the advantages of using HHO. No single meta-heuristic algorithm is suitable for the engineering applications used, so it is necessary to find a suitable metaheuristic algorithm in combination with specific application scenarios.

Meta-heuristic algorithms are mostly developed around physical phenomena, biological evolution, and group intelligence [11,12]. Among the many meta-heuristic algorithms, the most commonly used is the swarm intelligence optimization algorithm, such as the Harris Hawk optimization algorithm (HHO) [13], grey wolf optimizer (GWO) [14], butterfly optimization algorithm (BOA) [15], and so on. The sparrow search algorithm (SSA) [16] is a typical swarm intelligence optimization algorithm, which was proposed by Xue et al. in 2020. It has been widely used in many engineering fields, including UAV path planning [17], price prediction [18], and wind energy prediction [19]. In fact, the sparrow search algorithms suffer from the same problems as other meta-heuristic algorithms, such as low optimization-seeking accuracy and slow convergence. Thus, SSA needs to be improved. Yang et al. [20] used a chaotic mapping strategy to initialize the population and introduced adaptive weighting strategy and t-distribution variation strategy to balance the local exploration ability and global utilization ability of the algorithm. Yuan et al. [21] introduced centroid opposition learning, learning coefficients, and mutation operators in the original SSA to prevent SSA from falling into the local optima, and applied the improved algorithm to the control of distributed maximum power point tracking. Liu et al. [22] proposed a hybrid sparrow search algorithm based on constructing similarity, which overcomes the problem of the algorithm falling into a local optimum by introducing an improved chaotic mapping circle and t-distribution variation.

In summary, this paper compares the performance of 15 common meta-heuristic algorithms in node localization, and the comparison results are shown in Section 4. Then, SSA with better comprehensive performance is selected as the algorithm for the coordinate calculation stage, and SSA is improved with regard to its problems. Finally, the improved sparrow search algorithm (ISSA) is applied to the node localization of HWSNs. The main contributions of this paper are as follows:

- An improved sparrow search algorithm that incorporates the golden sine strategy, particle swarm optimal idea, and Gaussian perturbation is proposed. It shows a better performance in finding the optimal than the sparrow search algorithm and other comparative algorithms;
- ISSA is applied to the problem of solving the coordinates of unknown nodes in HWSNs. It achieves better localization accuracy compared with the localization algorithm using the remaining 15 meta-heuristic algorithms and LS.

The rest of this paper is organized as follows. In Section 2, the original SSA is introduced, and ISSA is presented to address the problems of slow convergence and low accuracy of SSA. In Section 3, simulation results are described and discussed. In Section 4, the application of ISSA in the node localization of HWSNs is introduced. Finally, Section 5 gives the summary of this paper and the direction of future work.

2. Basic SSA and Its Improvement

2.1. Basic SSA

SSA completes the spatial search by iteratively updating the position of each sparrow [16], and the entire population is divided into three categories: producer, scrounger, and investigator. The producers and scroungers each make up a certain proportion of the population and are dynamically updated according to the results of each iteration. However, the investigators are selected randomly from the whole population, usually at a rate of 10 to 20%. The details of the three categories are described as follows.

2.1.1. Updating Producer's Location

The producers are primarily responsible for searching for food and guiding the movement of the entire population. When the warning value R_2 is less than the safety value ST, it indicates that no predators were discovered during the search, necessitating a broad search by the producers. When the warning value R_2 exceeds the safe value ST, the sparrow has encountered a predator and is obliged to guide all scroungers to a safe area. The location update equation of the producers is shown in Equation (1).

$$X_{i,j}^{t+1} = \begin{cases} X_{i,j}^t \cdot \exp\left(\frac{-i}{\alpha \cdot iter_{\max}}\right) & \text{if } R_2 < ST \\ X_{i,j}^t + Q \cdot L & \text{if } R_2 \geq ST \end{cases} \tag{1}$$

where t represents the current iteration, i and j are the individual numbers of the population and the dimensions of the solution problem, respectively, $iter_{\max}$ is the maximum number of iterations. $\alpha \in (0,1]$ represents a random, and Q denotes a random number following the normal distribution. L is a matrix of $1 \times d$ in which each element is 1, where d represents the dimension of the solved problem. $X_{i,j}^t$ and $X_{i,j}^{t+1}$ are the position of the i-th sparrow's j-th dimension in the t-th and $(t + 1)$-th iterations, respectively. $R_2 \in (0,1]$ represents an alarm value, and $ST \in [0.5, 1)$ represents the safety threshold.

2.1.2. Updating Scrounger's Location

The remainder of the sparrow colony are scroungers whose primary function is to frequently monitor the producers. Once producers discover a source of good food, they will immediately abandon their current location in order to compete for it. Otherwise, the scroungers will go hungry and will be forced to fly to other locations in search of food. The location update equation of the scroungers is shown in Equation (2).

$$
X_{i,j}^{t+1} =
\begin{cases}
Q \cdot \exp\left(\dfrac{X_{worst}^t - X_{i,j}^t}{i^2}\right) & \text{if } i > n/2 \\[2mm]
X_p^{t+1} + \left|X_{i,j}^t - X_p^{t+1}\right| \cdot A^+ \cdot L & \text{otherwise}
\end{cases}
\tag{2}
$$

where X_{worst}^t represents the worst position in the t-th iteration and X_p^{t+1} denotes the optimal position of the producer in the $(t+1)$-th iteration. A is a matrix of $1 \times d$, in which the value of each element is randomly 1 or -1, and $A^+ = A^T(AA^T)^{-1}$. When $i > n/2$, it indicates that the sparrow is starving.

2.1.3. Updating Investigator's Location

The investigators are primarily in charge of colony security. When they detect danger, sparrows on the colony's periphery flee to a safe area. Those in the colony's center, on the other hand, walk at random. The location update equation of the investigators is shown in Equation (3).

$$
X_{i,j}^{t+1} =
\begin{cases}
X_{best}^t + \beta \cdot \left|X_{i,j}^t - X_{best}^t\right| & \text{if } f_i > f_g \\[2mm]
X_{i,j}^t + K\left(\dfrac{\left|X_{i,j}^t - X_{worst}^t\right|}{(f_i - f_w) + \varepsilon}\right) & \text{if } f_i = f_g
\end{cases}
\tag{3}
$$

where X_{best}^t represents the global optimal position in the t-th iteration, β denotes a normally distributed random number with mean 0 and variance 1, and k is a random number within $[-1, 1]$. f_i, f_w and f_g represent the fitness values of the i-th sparrow, the worst individual, and the best individual, respectively. $f_i > f_g$ means the sparrow is at the edge of the colony, and $f_i = f_g$ means the sparrow is in the middle of the colony.

According to Equation (1), the further back a producer is ranked in the fitness ranking, the more likely it is to fall into a local optimum. As a result, it is necessary to improve Equation (2) in order to improve the convergence accuracy of SSA. As shown in Equation (3), the vigilantes' location updates are random, which is detrimental to SSA's convergence speed.

2.2. Improved Sparrow Algorithm

2.2.1. Introduction of Golden Sine Strategy

The horizontal coordinates in Figure 1 represent producers sorted by fitness values, while the vertical coordinates represent the coefficient of variation of each producer's position. As illustrated in Figure 1, the larger the producer's ordinal number, the smaller the coefficient of variation of the position, which increases the probability of the producer falling into the local optimum. In summary, Equation (1) does not carry out a global search.

To improve the problem of the weak global search ability of the producers, the golden sine algorithm (Gold-SA) was introduced [23]. This algorithm traverses the whole search space by the relationship between the unit circle and the sine function on the one hand, and on the other hand, it gradually moves from the boundary of each dimension to the middle by the number of golden divisions until the best position for each dimension is found. Thus, in the paper, the golden sine algorithm is used to improve the global search ability of the producers in SSA. The equation of the discoverer position update after the introduction of the golden sine algorithm strategy is shown in Equation (4).

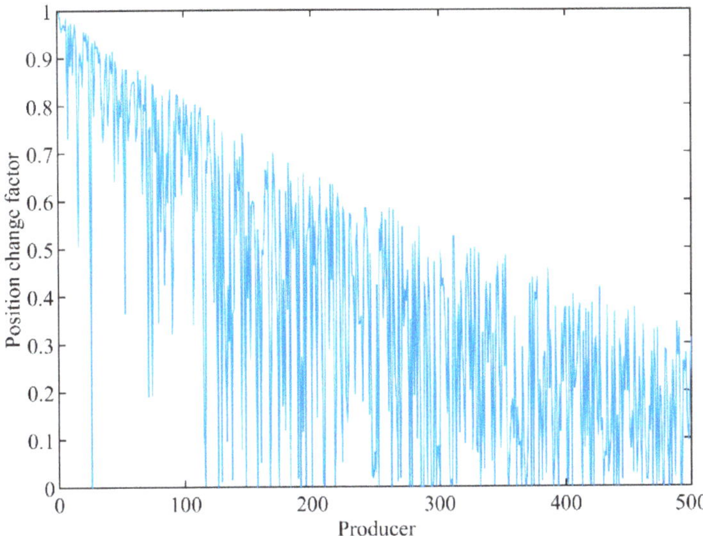

Figure 1. Relationship between location variation coefficient and numbering of producers.

$$X = \begin{cases} X_{b,j}^t \cdot |\sin(r_1)| + r_2 \cdot \sin(r_1) \cdot |x_1 \cdot X - x_2 \cdot X| & \text{if } R_2 < ST \\ X_{i,j}^t + Q \cdot L & \text{if } R_2 \geq ST \end{cases} \tag{4}$$

where r_1 and r_2 are the random numbers within $[0, 2\pi]$ and $[0, \pi]$, respectively. $X_{b,j}^t$ represents the position of the j-th dimension of the optimal individual in the t-th iteration. The initial value of x_1 is $a \cdot (1 - \tau) + b \cdot \tau$, and the initial value of x_2 is $a \cdot \tau + b \cdot (1 - \tau)$, where τ is the golden mean and the initial value of a and b is $-\pi$ and π, respectively. x_1, x_2, a, and b dynamically are updated as shown in Algorithm 1.

Algorithm 1: Pseudo-code for partial parameter update of Gold-SA

/* *F* represents the current fitness value, *G* represents the global optimal value,
    random$_1$ and *random*$_2$ represent random numbers between [0, 1] */
Input: $x_1 \leftarrow a \cdot (1 - \tau) + b \cdot \tau$; $x_2 \leftarrow a \cdot \tau + b \cdot (1 - \tau)$; $a \leftarrow -\pi$; $b \leftarrow \pi$;
Output: x_1, x_2
1:    if $(F < G)$ then
2:       $b \leftarrow x_2$; $x_2 \leftarrow x_1$; $x_1 \leftarrow a \cdot \tau + b \cdot (1 - \tau)$;
3:      else
4:       $a \leftarrow x_1$; $x_1 \leftarrow x_2$; $x_2 \leftarrow a \cdot (1 - \tau) + b \cdot \tau$;
5:      end if
6:    if $(x_1 = x_2)$ then
7:       $a \leftarrow random_1$; $b \leftarrow random_2$;
8:       $x_1 \leftarrow a \cdot \tau + b \cdot (1 - \tau)$; $x_2 \leftarrow a \cdot (1 - \tau) + b \cdot \tau$;
9:    end if

2.2.2. Introduction of Individual Optimal Strategies

By comparing the fitness values of individuals to the global optimal, the sparrow population investigators contribute to the coordination of global exploration and local search in the whole space search. Further, the idea of individual optimality in the particle swarm algorithm can strengthen the convergence ability of the algorithm [24], and the investigators' position update formula after introducing the optimal individual is as in Equation (5).

$$X_{i,j}^{t+1} = \begin{cases} X_{best}^t + \beta \cdot \left| X_{pi,j}^t - X_{best}^t \right| & \text{if } f_i > f_g \\ X_{pi,j}^t + K \left(\dfrac{\left| X_{pi,j}^t - X_{worst}^t \right|}{(f_{pi} - f_w) + \varepsilon} \right) & \text{if } f_i = f_g \end{cases} \tag{5}$$

where $X_{pi,j}^t$ represents the j-th dimensional value of the i-th sparrow's historical optimal individual.

2.2.3. Gaussian Perturbation

The sparrow algorithm, like the majority of swarm intelligence optimization algorithms, suffer from the problem of falling into a local optimum, which results in poor searching accuracy. To address this issue, it is possible to use the fitness value of the population optimal individual to determine whether the algorithm is trapped in a local optimum [25]. If the global optimal individual's fitness value is less than a threshold in two consecutive iterations, the algorithm is said to be trapped in a local optimum. At this time, a Gaussian random wandering strategy can be introduced to apply perturbation to the optimal position to help the algorithm to jump out of the local optimum. The improved way of updating the position of the investigators is shown in Equations (6)–(8).

$$\hat{X}_t = Gaussian(G_t, \sigma) \tag{6}$$

$$\sigma = \cos\left(\pi/2 \cdot (t/T)^2 \right) \cdot (G_t - X_t^*) \tag{7}$$

$$G_t = \begin{cases} \hat{X}_t, & f(\overline{X}_t) < f(G_t) \\ G_t, & otherwise \end{cases} \tag{8}$$

where X_t is the position of the optimal individual after applying Gaussian perturbation at the t-th iteration, and X_t^* is a random individual of the population.

3. Experimental Results and Analysis

HHO [13], SSA [16], PSO [24], Gold-SA [25], and artificial gorilla troops optimizer (GTO) [26] are selected as the comparison algorithms to verify the search performance of ISSA. The population size of all algorithms is set to 30, and the number of iterations is set to 500. In the ISSA proposed in this paper, the ratio of generators is 0.6, the ratio of discoverers is 0.7, and the ratio of vigilantes is 0.2. Meanwhile, Gaussian perturbation is introduced when the global optimum value is below the threshold 1.00×10^{-10} in two consecutive iterations. The parameters of the other algorithms are consistent with their literature. The whole experiment is divided into four parts: convergence accuracy analysis, convergence speed analysis, rank sum test, and complexity analysis. The simulation experiment described in this article was conducted on a Windows 11 64-bit operating system with an AMD Ryzen 7 5800H processor with Radeon Graphics 3.20 GHz, 16 GB of RAM, and MATLAB 2014B. All experimental results are the average values after 30 runs.

In this section, 23 test functions are selected from CEC's benchmark functions, which are commonly used for optimization testing of meta-heuristic algorithms [16,27–29]. The functions F_1–F_7 are unimodal functions, the functions F_8–F_{13} are multimodal functions, and the functions F_{14}–F_{23} are fixed-dimension functions. Table 1 lists the parameters of each function, including its expression, dimension, range, and optimal value.

Table 1. Benchmark function.

Name	Formula of Functions	Dim	Range	Best
Sphere	$F_1(x) = \sum_{i=1}^{n} x_i^2$	30	[100, 100]	0
Schwefel 2.22	$F_2(x) = \sum_{i=1}^{n} |x_i| + \prod_{i=1}^{n} |x_i|$	30	[10, 10]	0
Schwefel 1.2	$F_3(x) = \sum_{i=1}^{n} (\sum_{j-1}^{i} x_j)^2$	30	[100, 100]	0
Schwefel 2.21	$F_4(x) = \max_i\{|x_i|, 1 \leq i \leq n\}$	30	[100, 100]	0
Rosenbrock	$F_5(x) = \sum_{i=1}^{n-1} [100(x_{i+1} - x_i^2)^2 + (x_i - 1)^2]$	30	[30, 30]	0
Step	$F_6(x) = \sum_{i=1}^{n} ([x_i + 0.5])^2$	30	[100, 100]	0
Quartic	$F_7(x) = ix_i^4 + random[0, 1]$	30	[128, 128]	0
Schwefel 2.26	$F_8(x) = \sum_{i=1}^{n} - x_i \sin(\sqrt{|x_i|})$	30	[500, 500]	$-418.9829 \times n$
Rastrigrin	$F_9(x) = \sum_{i=1}^{n} [x_i^2 - 10\cos(2\pi x_i) + 10]$	30	[5.12, 5.12]	0
Ackley	$F_{10}(x) = -20\exp\left(-0.2\sqrt{\frac{1}{n}\sum_{i=1}^{n} x_i^2}\right) - \exp\left(\frac{1}{n}\sum_{i=1}^{n} \cos(2\pi x_i)\right) + 20 + e$	30	[32, 32]	0
Griewank	$F_{11}(x) = \frac{1}{4000}\sum_{i=1}^{n} x_i^2 - \prod_{i=1}^{n} \cos(\frac{x_i}{\sqrt{i}}) + 1$	30	[600, 600]	0
Penalized 1	$F_{12}(x) =$ $\frac{\pi}{n}\{10\sin^2(\pi y_1) + \sum_{i=1}^{n-1} (y_i - 1)^2[1 + 10\sin^2(\pi y_{i+1})] + (y_{n-1})^2\} + \sum_{i=1}^{n} u(x_i, 10, 100, 4)$ $y_i = 1 + \frac{x_i+1}{4} \quad u = (x_i, a, k, m) = \begin{cases} k(x_i - a)^m, & x_i > a \\ 0, & -a < x_i < a \\ k(-x_i - a)^m, & x_i < a \end{cases}$	30	[50, 50]	0
Penalized 2	$F_{13}(x) = 0.1\{\sin^2(3\pi x_i) + \sum_{i=1}^{n} (x_i - 1)^2[1 + \sin^2(3\pi x_{i+1})] + (x_i - 1)^2(1 + \sin^2(2\pi x_{i+1}))\}$ $+ \sum_{i=1}^{n} u(x_i, 5, 100, 4)$	30	[50, 50]	0
Foxholes	$F_{14}(x) = (\frac{1}{500} + \sum_{j=1}^{25} \frac{1}{j+\sum_{i=1}^{2} (x_i - a_{ij})^6})^{-1}$	2	[−65, 65]	1
Kowalik	$F_{15}(x) = \sum_{i=1}^{11} [a_i - x_1(b_i^2 + b_i x_2)/(b_i^2 + b_i x_3 + x_4)]^2$	4	[−5, −5]	0.00030
Six-Hump Gamel	$F_{16}(x) = 4x_1^2 - 2.1x_1^4 + \frac{1}{3}x_1^6 + x_1 x_2 - 4x_2^2 + 4x_2^4$	2	[−5, −5]	−1.0316
Branin	$F_{17}(x) = (x_2 - \frac{5.1}{4\pi^2} x_1^2 + \frac{5}{\pi} x_1 - 6)^2 + 10(1 - \frac{1}{8\pi}) \cos x_1$	2	[−5, −5]	0.398
Goldstein-price	$F_{18}(x) = [1 + (x_1 + x_2 + 1)^2(19 - 14x_1 + 3x_1^2 - 14x_2 + 6x_1 x_2 + 3x_2^2)]$ $*[30 + (2x_1 - 3x_2)^2 * (18 - 32x_1 + 12x_1^2 + 48x_2 - 36x_1 x_2 + 27x_2^2)]$	2	[−2, 2]	3

Table 1. *Cont.*

Name	Formula of Functions	Dim	Range	Best
Hartmann 3-D	$F_{19}(x) = -\sum_{i=1}^{4} c_i \exp\left(-\sum_{j=1}^{3} a_{ij}(x_i - p_{ij})^2\right)$	3	[1, 3]	−3.86
Hartmann 6-D	$F_{20}(x) = -\sum_{i=1}^{4} c_i \exp\left(-\sum_{j=1}^{6} a_{ij}(x_i - p_{ij})^2\right)$	6	[0, 1]	−3.32
Shekel 1	$F_{21}(x) = -\sum_{i=1}^{5} \left[(X - a_i)(X - a_i)^T + c_i\right]^{-1}$	4	[0, 10]	−10.1532
Shekel 2	$F_{22}(x) = -\sum_{i=1}^{7} \left[(X - a_i)(X - a_i)^T + c_i\right]^{-1}$	4	[0, 10]	−10.4029
Shekel 3	$F_{23}(x) = -\sum_{i=1}^{10} \left[(X - a_i)(X - a_i)^T + c_i\right]^{-1}$	4	[0, 10]	−10.5364

3.1. Convergence Accuracy Analysis

Table 2 presents the average (AVG) and standard deviation (STD) of the average localization errors for the six meta-heuristics. Among them, R in Table 2 is explained in detail in Section 3.2.

Table 2. Comparison of benchmark function results.

Function		PSO	Gold-SA	HHO	GTO	SSA	ISSA
F_1	AVG	8.35×10^{-6}	4.46×10^{-207}	1.10×10^{-98}	**0.0**	1.11×10^{-84}	**0.0**
	STD	8.80×10^{-6}	0.0	4.65×10^{-98}	**0.0**	5.85×10^{-84}	**0.0**
	p	$4.13 \times 10^{-11}(+)$	$9.64 \times 10^{-2}(-)$	$1.07 \times 10^{-9}(+)$	$7.27 \times 10^{-8}(+)$	$7.27 \times 10^{-8}(+)$	
F_2	AVG	1.12×10^{-2}	1.69×10^{-129}	8.34×10^{-51}	4.06×10^{-193}	5.25×10^{-55}	**0.0**
	STD	1.79×10^{-2}	9.10×10^{-129}	3.54×10^{-50}	0.0	2.78×10^{-54}	**0.0**
	p	$4.46 \times 10^{-11}(+)$	$5.14 \times 10^{-2}(-)$	$2.45 \times 10^{-10}(+)$	$3.19 \times 10^{-3}(+)$	$2.24 \times 10^{-11}(+)$	
F_3	AVG	$4.13 \times 10^{+02}$	1.47×10^{-206}	4.36×10^{-72}	**0.0**	5.32×10^{-82}	**0.0**
	STD	$1.28 \times 10^{+03}$	0.0	2.34×10^{-71}	**0.0**	2.82×10^{-81}	**0.0**
	p	$4.01 \times 10^{-11}(+)$	$7.42 \times 10^{-2}(-)$	$6.48 \times 10^{-10}(+)$	$2.01 \times 10^{-07}(+)$	$2.01 \times 10^{-07}(+)$	
F_4	AVG	3.71×10^{-01}	7.87×10^{-96}	5.22×10^{-49}	3.71×10^{-194}	8.22×10^{-46}	**0.0**
	STD	1.16×10^{-01}	4.24×10^{-95}	2.65×10^{-48}	0.0	4.40×10^{-45}	**0.0**
	p	$3.00 \times 10^{-11}(+)$	$1.67 \times 10^{-1}(-)$	$1.69 \times 10^{-8}(+)$	$4.97 \times 10^{-4}(+)$	$1.93 \times 10^{-10}(+)$	
F_5	AVG	2.87×10^{1}	6.62×10^{-3}	1.45×10^{-2}	1.61	1.70×10^{-4}	$\mathbf{1.41 \times 10^{-8}}$
	STD	9.06	1.01×10^{-02}	1.82×10^{-02}	6.01	4.18×10^{-4}	$\mathbf{3.71 \times 10^{-8}}$
	p	$3.02 \times 10^{-11}(+)$	$2.38 \times 10^{-07}(+)$	$8.15 \times 10^{-11}(+)$	$1.52 \times 10^{-3}(+)$	$6.07 \times 10^{-11}(+)$	
F_6	AVG	5.02×10^{-6}	2.31×10^{-4}	1.37×10^{-4}	1.77×10^{-7}	3.90×10^{-7}	$\mathbf{2.46 \times 10^{-11}}$
	STD	5.90×10^{-6}	3.07×10^{-4}	2.19×10^{-4}	1.80×10^{-7}	5.04×10^{-7}	$\mathbf{7.98 \times 10^{-11}}$
	p	$2.83 \times 10^{-8}(+)$	$5.49 \times 10^{-11}(+)$	$8.15 \times 10^{-11}(+)$	$0.379(-)$	$4.50 \times 10^{-11}(+)$	
F_7	AVG	7.57×10^{-2}	2.16×10^{-4}	1.73×10^{-4}	9.01×10^{-5}	3.67×10^{-4}	$\mathbf{6.57 \times 10^{-5}}$
	STD	2.68×10^{-2}	2.84×10^{-4}	2.24×10^{-4}	8.64×10^{-5}	3.27×10^{-4}	$\mathbf{4.81 \times 10^{-5}}$
	p	$3.02 \times 10^{-11}(+)$	$1.44 \times 10^{-3}(+)$	$4.94 \times 10^{-5}(+)$	$9.06 \times 10^{-8}(+)$	$4.20 \times 10^{-10}(+)$	
F_8	AVG	-2.70×10^{3}	-1.26×10^{4}	-1.26×10^{4}	-1.26×10^{4}	-9.34×10^{3}	$\mathbf{-1.26 \times 10^{4}}$
	STD	3.74×10^{2}	1.61×10^{-1}	5.96×10^{-1}	7.49×10^{-5}	2.49×10^{3}	$\mathbf{2.90 \times 10^{-8}}$
	p	$3.02 \times 10^{-11}(+)$	$4.62 \times 10^{-10}(+)$	$1.55 \times 10^{-9}(+)$	$3.02 \times 10^{-11}(+)$	$2.91 \times 10^{-11}(+)$	
F_9	AVG	52.1	0.0	0.0	0.0	0.0	0.0
	STD	12.2	0.0	0.0	0.0	0.0	0.0
	p	$1.21 \times 10^{-12}(+)$	NaN(=)	NaN(=)	NaN(=)	NaN(=)	
F_{10}	AVG	1.74×10^{-3}	8.88×10^{-16}	8.88×10^{-16}	8.88×10^{-16}	8.88×10^{-16}	8.88×10^{-16}
	STD	1.18×10^{-3}	9.86×10^{-32}	9.86×10^{-32}	9.86×10^{-32}	9.86×10^{-32}	9.86×10^{-32}
	p	$1.21 \times 10^{-12}(+)$	NaN(=)	NaN(=)	NaN(=)	NaN(=)	
F_{11}	AVG	42.0	0.0	0.0	0.0	0.0	0.0
	STD	5.85	0.0	0.0	0.0	0.0	0.0
	p	$1.21 \times 10^{-12}(+)$	NaN(=)	NaN(=)	NaN(=)	NaN(=)	
F_{12}	AVG	8.00×10^{-1}	1.53×10^{-5}	1.15×10^{-5}	2.68×10^{-8}	3.80×10^{-8}	$\mathbf{7.50 \times 10^{-11}}$
	STD	9.05×10^{-1}	2.77×10^{-5}	1.78×10^{-5}	4.94×10^{-8}	5.16×10^{-8}	$\mathbf{2.47 \times 10^{-10}}$
	p	$3.02 \times 10^{-11}(+)$	$2.67 \times 10^{-9}(+)$	$8.10 \times 10^{-10}(+)$	$0.3871(-)$	$3.16 \times 10^{-10}(+)$	
F_{13}	AVG	1.10×10^{-3}	5.83×10^{-5}	1.54×10^{-4}	2.93×10^{-3}	6.16×10^{-7}	$\mathbf{4.19 \times 10^{-11}}$
	STD	3.30×10^{-3}	1.23×10^{-4}	2.16×10^{-4}	8.48×10^{-3}	7.00×10^{-7}	$\mathbf{1.14 \times 10^{-10}}$
	p	$3.87 \times 10^{-1}(-)$	$3.32 \times 10^{-6}(+)$	$4.20 \times 10^{-10}(+)$	$0.3555(-)$	$6.07 \times 10^{-11}(+)$	
F_{14}	AVG	1.30	1.03	1.29	$\mathbf{9.98 \times 10^{-1}}$	8.73	4.76
	STD	5.82×10^{-1}	1.79×10^{-1}	9.24×10^{-1}	$\mathbf{3.33 \times 10^{-16}}$	4.98	5.34
	p	$7.44 \times 10^{-10}(+)$	$9.69 \times 10^{-7}(+)$	$2.05 \times 10^{-6}(+)$	$6.12 \times 10^{-13}(+)$	$4.20 \times 10^{-5}(+)$	
F_{15}	AVG	4.75×10^{-4}	4.00×10^{-4}	3.50×10^{-4}	3.99×10^{-4}	3.21×10^{-4}	$\mathbf{3.08 \times 10^{-4}}$
	STD	2.76×10^{-4}	2.42×10^{-4}	3.20×10^{-5}	2.75×10^{-4}	2.55×10^{-5}	$\mathbf{7.64 \times 10^{-7}}$
	p	$8.31 \times 10^{-3}(+)$	$1.09 \times 10^{-5}(+)$	$1.64 \times 10^{-5}(+)$	$7.64 \times 10^{-8}(+)$	$3.82 \times 10^{-10}(+)$	

Table 2. *Cont.*

Function		PSO	Gold-SA	HHO	GTO	SSA	ISSA
F_{16}	AVG	**−1.03**	−1.03	−1.03	**−1.03**	−1.03	−1.03
	STD	**0.0**	4.05×10^{-3}	1.23×10^{-8}	**0.0**	1.16×10^{-8}	**0.0**
	p	$1.21 \times 10^{-12}(+)$	$3.02 \times 10^{-11}(+)$	$8.88 \times 10^{-1}(-)$	$1.21 \times 10^{-12}(+)$	$1.21 \times 10^{-12}(+)$	
F_{17}	AVG	$\mathbf{3.98 \times 10^{-1}}$	4.00×10^{-1}	3.98×10^{-1}	$\mathbf{3.98 \times 10^{-1}}$	3.98×10^{-1}	3.98×10^{-1}
	STD	$\mathbf{1.11 \times 10^{-16}}$	1.28×10^{-2}	4.06×10^{-5}	$\mathbf{1.11 \times 10^{-16}}$	1.32×10^{-8}	3.72×10^{-16}
	p	$1.21 \times 10^{-12}(+)$	$3.02 \times 10^{-11}(+)$	$2.77 \times 10^{-5}(+)$	$1.21 \times 10^{-12}(+)$	$1.72 \times 10^{-12}(+)$	
F_{18}	AVG	3.0	14.2	3.0	**3.0**	3.0	3.0
	STD	3.96×10^{-15}	13.5	5.54×10^{-7}	$\mathbf{1.78 \times 10^{-15}}$	1.34×10^{-8}	3.35×10^{-15}
	p	$6.32 \times 10^{-12}(+)$	$3.02 \times 10^{-11}(+)$	$6.63 \times 10^{-1}(-)$	$1.72 \times 10^{-12}(+)$	$4.08 \times 10^{-12}(+)$	
F_{19}	AVG	**−3.86**	−3.8	−3.86	**−3.86**	−3.0	**−3.86**
	STD	$\mathbf{2.66 \times 10^{-15}}$	8.01×10^{-2}	3.04×10^{-3}	$\mathbf{2.66 \times 10^{-15}}$	5.07×10^{-5}	$\mathbf{2.66 \times 10^{-15}}$
	p	$1.21 \times 10^{-12}(+)$	$3.34 \times 10^{-11}(+)$	$3.20 \times 10^{-9}(+)$	$1.21 \times 10^{-12}(+)$	$1.21 \times 10^{-12}(+)$	
F_{20}	AVG	−3.25	−2.95	−3.1	−3.26	−3.26	**−3.27**
	STD	5.89×10^{-2}	3.71×10^{-1}	9.34×10^{-2}	5.94×10^{-2}	8.00×10^{-2}	$\mathbf{5.89 \times 10^{-2}}$
	p	$1.58 \times 10^{-2}(+)$	$9.83 \times 10^{-8}(+)$	$3.52 \times 10^{-7}(+)$	$3.485 \times 10^{-3}(+)$	$8.12 \times 10^{-4}(+)$	
F_{21}	AVG	−6.31	−10.2	−5.19	**−10.2**	−10.2	**−10.2**
	STD	3.46	5.63×10^{-3}	7.46×10^{-1}	$\mathbf{1.78 \times 10^{-15}}$	1.12×10^{-5}	$\mathbf{1.78 \times 10^{-15}}$
	p	$5.00 \times 10^{-1}(-)$	$1.86 \times 10^{-9}(+)$	$3.02 \times 10^{-11}(+)$	$1.21 \times 10^{-12}(+)$	$1.21 \times 10^{-12}(+)$	
F_{22}	AVG	−6.61	−10.4	−5.73	**−10.4**	−10.4	−10.4
	STD	3.8	5.11×10^{-3}	1.67	**0.0**	3.02×10^{-4}	**0.0**
	p	$1(-)$	$1.96 \times 10^{-10}(+)$	$3.02 \times 10^{-11}(+)$	$1.21 \times 10^{-12}(+)$	$1.21 \times 10^{-12}(+)$	
F_{23}	AVG	−6.81	−10.5	−5.05	−10.5	−10.5	−10.5
	STD	3.78	3.19×10^{-3}	4.18×10^{-1}	1.93×10^{-14}	5.71×10^{-6}	$\mathbf{8.88 \times 10^{-15}}$
	p	$1(-)$	$3.02 \times 10^{-11}(+)$	$3.02 \times 10^{-11}(+)$	$1.72 \times 10^{-12}(+)$	$1.21 \times 10^{-12}(+)$	

According to Table 2, ISSA can find the optimal value in the unimodal functions $F_1 \sim F_4$. This indicates that the introduction of the golden sine strategy significantly enhances the global optimality finding ability of the original SSSA. For the unimodal functions $F_5 \sim F_7$, although the optimal values are not sought, their convergence accuracy is higher than that of the remaining four compared algorithms. For the multimodal functions F_8, although the mean value of ISSA is the same as Gold-SA, HHO, and GTO, the standard deviation of ISSA is much lower than that of Gold-SA, HHO, and GTO. Among the multimodal functions $F_9 \sim F_{11}$, the convergence accuracy of ISSA is the same as that of Gold-SA, HHO, GTO, and SSA, and they can all converge to the optimal value. However, the convergence accuracy of ISSA is the best in all cases in the multimodal functions F_{12} and F_{13}. For the fixed-dimension functions $F_{14} \sim F_{23}$, a total of seven best or tied best convergence accuracies were obtained for ISSA.

The ranking results of the six meta-heuristic algorithms are shown in Figure 2. The ranking is determined primarily by the average value of each algorithm; if the average value is identical, the ranking is determined by the standard deviation.

The radar plot shown in Figure 2 has 23 polar axes, and each polar axis represents 1 test function. Starting from the center point of the radar plot, there are six circles that expand outward in sequence, and the numbers on the circles represent the ranking of the algorithms. A closed polygon can be obtained by connecting the ranking points of each algorithm on the 23 test functions. The smaller the area of the polygon, the better the convergence performance of the algorithm. According to Figure 2, it can be seen that the area enclosed by ISSA is the smallest, which indicates that the convergence accuracy of ISSA is better than that of PSO, Gold-SA, HHO, GTO, and SSA.

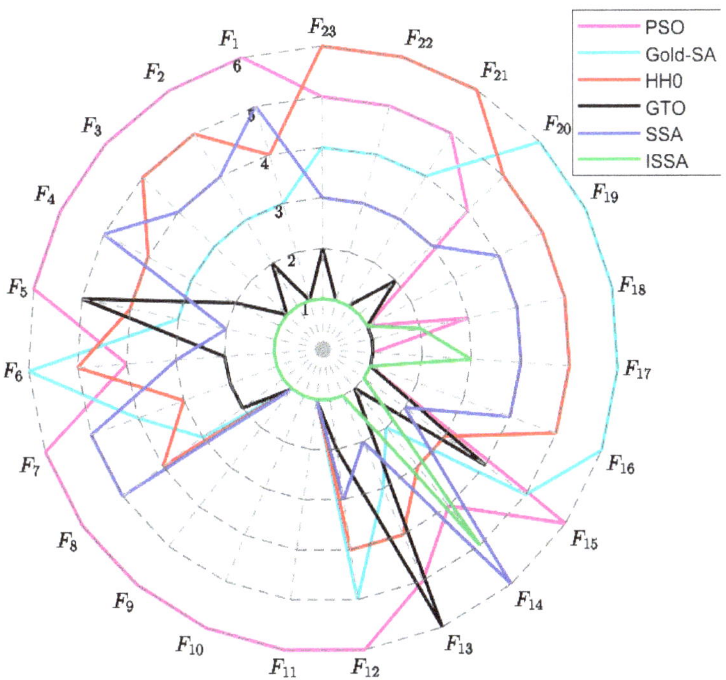

Figure 2. Radar chart of sort of algorithm.

3.2. Convergence Speed Analysis

The convergence curves for the six meta-heuristic algorithms on the 23 benchmark functions are depicted in Figure 3.

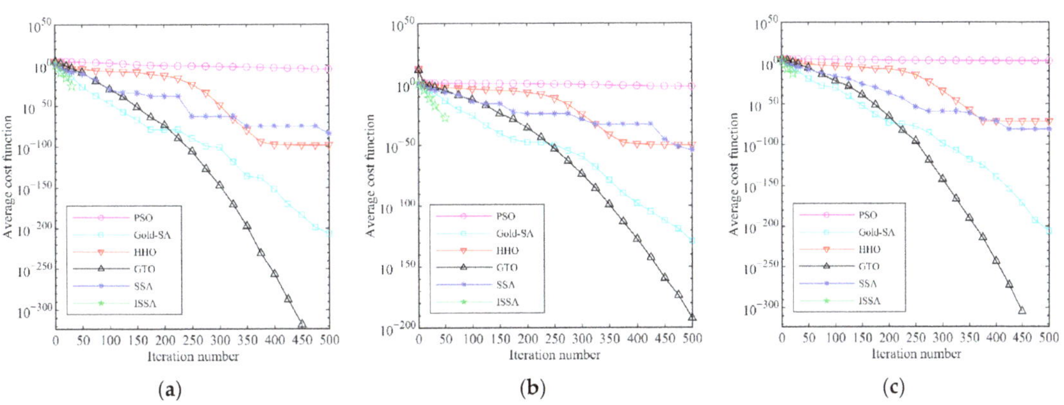

Figure 3. *Cont.*

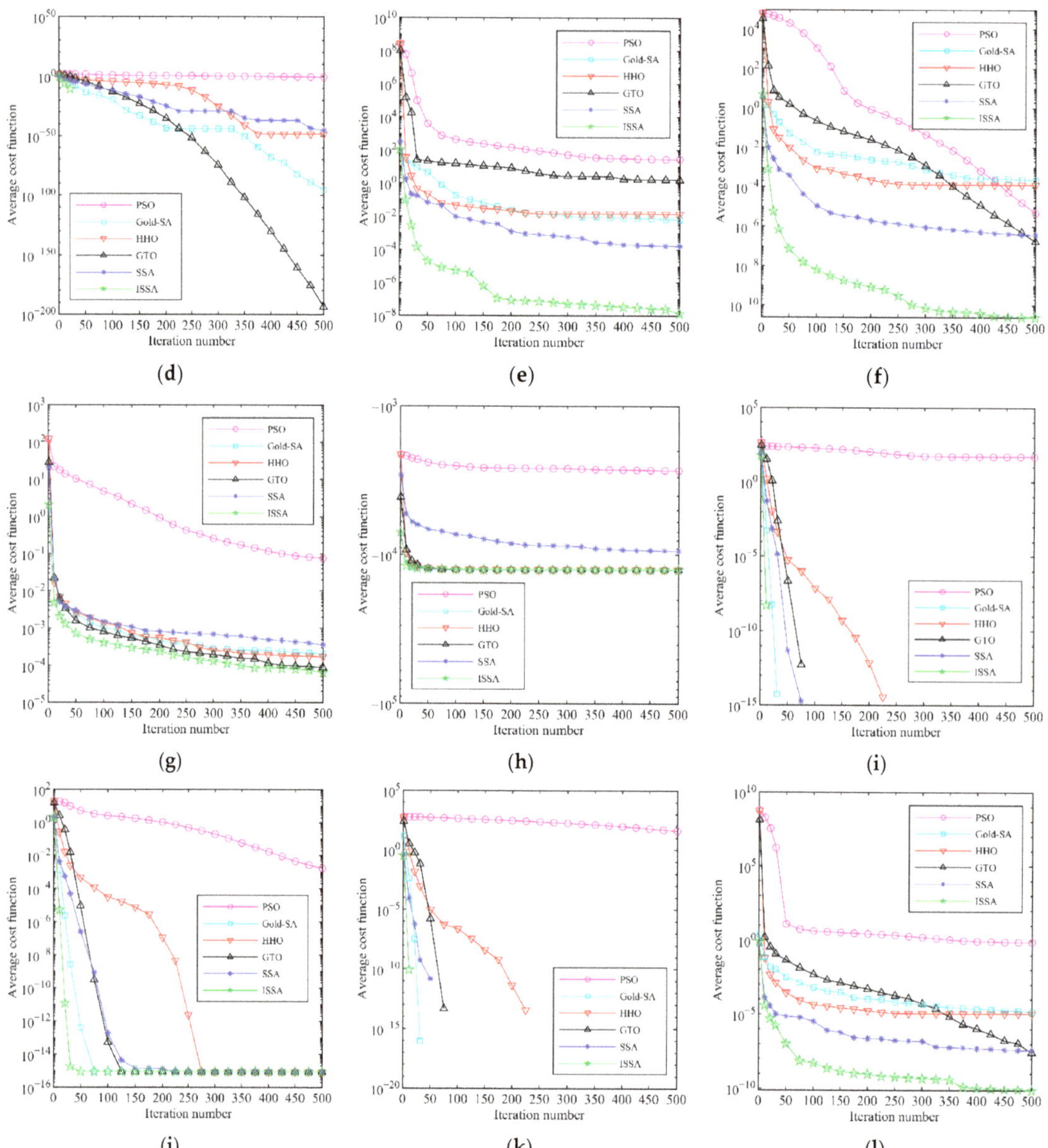

Figure 3. *Cont.*

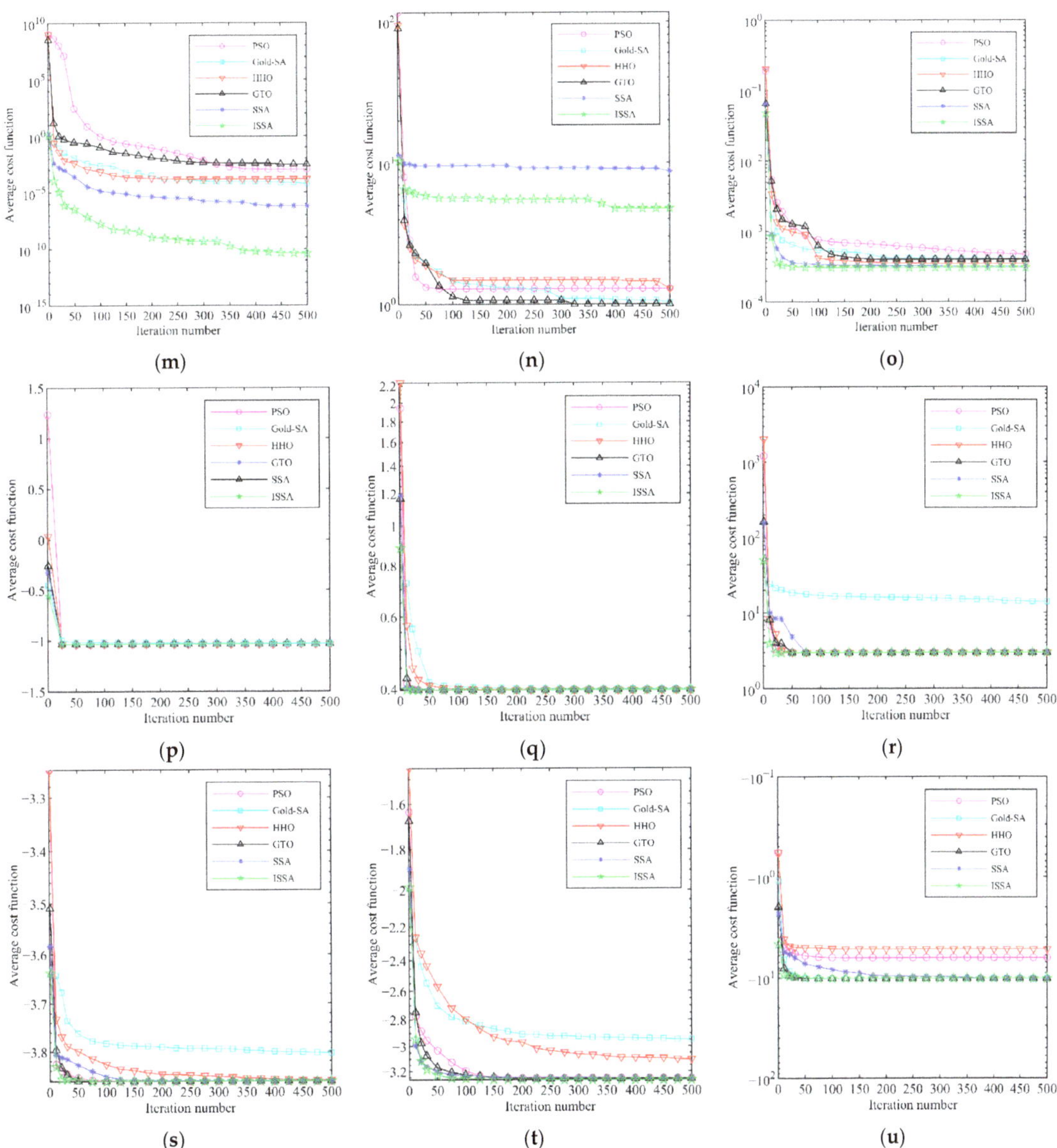

Figure 3. *Cont.*

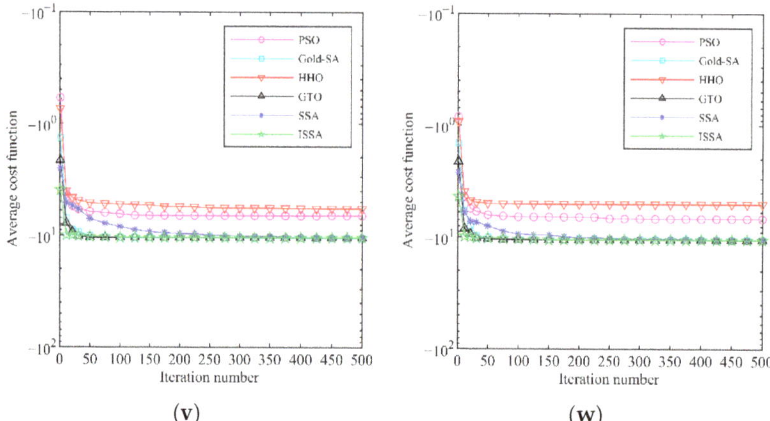

(v) (w)

Figure 3. Convergence curve of benchmark function: (**a**) F_1: Sphere; (**b**) F_2: Schwefel 2.22; (**c**) F_3: Schwefel 1.2; (**d**) F_4: Schwefel 2.21; (**e**) F_5: Rosenbrock; (**f**) F_6: Step; (**g**) F_7: Quartic; (**h**) F_8: Schwefel 2.26; (**i**) F_9: Rastrigrin; (**j**) F_{10}: Ackley; (**k**) F_{11}: Griewank; (**l**) F_{12}: Penalized 1; (**m**) F_{13}: Penalized 2; (**n**) F_{14}: Foxholes; (**o**) F_{15}: Kowalik; (**p**) F_{16}: Six-Hump Gamel; (**q**) F_{17}: Branin; (**r**) F_{18}: Goldstein-price; (**s**) F_{19}: Hartmann 3-D; (**t**) F_{20}: Hartmann 6-D; (**u**) F_{21}: Shekel 1; (**v**) F_{22}: Shekel 2; (**w**) F_{23}: Shekel 3.

As shown in Figure 3, the convergence speed of ISSA in the unimodal functions F_1~F_7 is faster than that of the remaining five compared algorithms. In particular, ISSA can converge to the optimal value within 60 times in F_1~F_4. This indicates that the introduction of the individual optimal idea of PSO in the position update of the investigator greatly accelerates the convergence speed of the algorithm. For the multimodal functions F_8~F_{13}, the convergence speed of ISSA is also better than the remaining five comparison algorithms. Among the fixed-dimensional functions F_{14}~F_{23}, except for F_{14}, the convergence rate of ISSA is not weaker than that of PSO, Gold-SA, HHO, GTO, and SSA. In summary, the improved ISSA has a faster convergence speed.

3.3. Wilcoxon Rank Sum Test

García et al. [30] proposed that it is not sufficient to evaluate the performance of meta-heuristic algorithms using only the average and standard deviation. Therefore, it is essential to perform statistical tests. The Wilcoxon rank sum test, a binary hypothesis, determines whether there is a significant difference between two samples. It is assumed that H_0: the overall difference between the two samples is not significant, and H_1: there is a significant difference between the two samples. If the significance level between two samples is less than 0.05, then the H_0 hypothesis can be rejected; otherwise, the H_1 hypothesis is rejected. The significance level between such algorithms is denoted by p.

In this section, the results of one run of each meta-heuristic algorithm are taken as 1 element in the corresponding sample, and there are 30 elements in each set of samples. Combining the sample data of ISSA and the sample data of either metaheuristic algorithm can help to find the p value between them. NaN implies that the two samples are identical and cannot be tested with the Wilcoxon rank sum test, which indicates that the performance of the two sets of algorithms is the same. The p values on each benchmark function are shown in Table 2. The + located in parentheses after p represents a significant difference between the two samples, and − represents no significant difference. p combined with the AVG corresponding to the two algorithms can effectively determine the superiority of the algorithms. If AVG of ISSA is smaller than AVG of the comparison algorithm and H_1 holds, it means that ISSA performs better than the comparison algorithm. If AVG of ISSA is larger than AVG of the comparison algorithm and H_1 holds, the performance of ISSA is weaker than that of the comparison algorithm. If AVG of ISSA is the same as AVG of the comparison algorithm and H_1 holds, then the two algorithms have the same performance. In the remaining cases, it can be regarded as uncertain in terms of statistical significance.

Table 3 gives the statistical results of the performance advantages and disadvantages between ISSA and the remaining five algorithms on 23 benchmark functions, respectively. As shown in Table 3, the performance of ISSA is better than that of SSA on 13 benchmark test functions, while the performance of ISSA is the same as that of SSA on the remaining 10 benchmark functions. This indicates that the proposed algorithm in this paper has achieved some success in terms of the performance of the optimization search. Compared with the GTO proposed by Abdollahzadeh et al. in 2021 [31], ISSA outperforms each other on all six benchmark functions and has the same performance on 13 benchmark functions. This indicates that ISSA still has some advantages over the recently proposed meta-heuristic algorithm.

Table 3. Statistical results of the performance advantages and disadvantages between ISSA and each algorithm.

ISSA and PSO	ISSA and Gold-SA	ISSA and HHO	ISSA and GTO	ISSA and SSA
14/1/4/4	10/1/8/4	13/1/7/2	6/1/13/3	13/0/10/0

a/b/c/d a, b, c denote the number of benchmark functions whose ISSA performance is better than, weaker than, or equal to the comparison algorithm, respectively. d denotes the number of benchmark functions that cannot be judged in terms of statistical significance.

3.4. Time Complexity Analysis

Let the population size of SSA and ISSA be N, the number of iterations be T_{max}, and the problem dimension be D. From Section 2, two algorithms differ only in the way they update different species of sparrows, and they have the same algorithmic complexity in the steps of population initialization, boundary check, position update, and sort update, all of which be $o(T_{max} \cdot N \cdot D)$. Assuming that the proportions of producers, scroungers, and investigators in the sparrow population are r_1, r_2, and r_3, respectively, then in SSA, the time complexity of the location update of producers is $o(T_{max} \cdot r_1 \cdot N \cdot D)$, the time complexity of the location update of scroungers is $o(T_{max} \cdot r_2 \cdot N \cdot D)$, and the location update of investigators is $o(T_{max} \cdot N \cdot D)$. The total time complexity of SSA is $o(T_{max} \cdot N \cdot D)$.

In ISSA, the time complexity of the position update of the producer after introducing the golden sine strategy is $o(T_{max} \cdot r_1 \cdot N^2)$, and the position update of the investigators after introducing the particle swarm optimal idea is still $o(T_{max} \cdot r_3 \cdot N)$. In addition, the time complexity added by introducing Gaussian perturbation to the optimal sparrow individuals is $o(T_{max} \cdot D)$. The total time complexity of ISSA is $o(T_{max} \cdot N \cdot D \cdot (1 + r_1 \cdot N))$.

In summary, the time complexity of ISSA is slightly higher than that of SSA for the same number of iterations. However, the convergence speed of ISSA is fast, and the time consumption of ISSA may be reduced if no more changes in the search results are taken as the end condition. ISSA is more suitable for engineering applications that require high search accuracy and have lower requirements for computation time.

4. Application of ISSA in Node Location in HWSNs

4.1. Node Localization Problem in HWSNs

The node localization of HWSNs can be divided into two stages: distance estimation and coordinate calculation. Once the distances from each anchor point to the unknown node are obtained in the distance estimation phase, the coordinate calculation of the unknown node can be transformed into an optimization problem, as shown in Equation (9). The dimension of this optimization problem is 2, and the theoretical optimal value is 0. Therefore, the problem can be solved by meta-heuristic algorithms.

$$F(x) = \sum_{j=1}^{Na} \left| \sqrt{(x_1 - a_j)^2 + (x_2 - b_j)^2} - d_j \right| \tag{9}$$

where Na indicates the number of anchor nodes, (a_j, b_j) represents the coordinates of the j-th unknown node, and d_j is the distance from the unknown node to the j-th anchor node.

4.2. Network Model

Figure 4a shows the network connection of 50 nodes in an area of 100×100 m^2, in which the number of anchor nodes represented by red $\bigcirc$ is 20, the number of unknown nodes represented by blue $\triangle$ is 30, and the blue solid line represents the neighboring nodes can communicate with each other. In addition, the communication radius of each node is different. The node localization of HWSNs is to find out the coordinates of unknown nodes using the information of finite anchor nodes, such as the coordinates of each anchor point and the distance from each anchor point to the unknown node. The distance from each anchor node to the unknown node is shown in Figure 4b, in which blue dashed line indicates the distance from the anchor node to the unknown node.

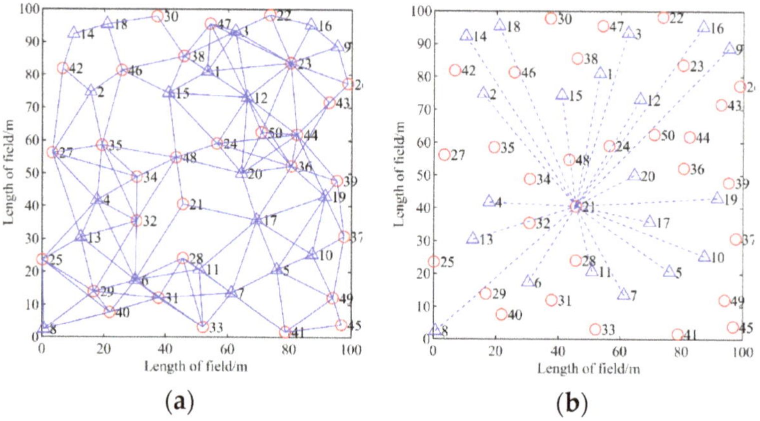

Figure 4. Heterogeneous wireless sensor network model: (**a**) network connections; (**b**) distance from all anchor nodes to unknown node 21.

4.3. Localization Steps

In this section, we replace the least-squares method of the localization algorithm with the meta-heuristic algorithm. The specific steps of the localization algorithm are as follows.

Step 1. Calculating the intersection area of the incoming neighbor

As shown in Figure 5a, when the intersection region of anchor nodes 1, 2, and 3 is used as the search space for meta-heuristic algorithms, the searching range can be narrowed, thus accelerating the convergence speed. However, because the intersection region of anchor points is an irregular graph, performing the calculation is difficult. To simplify computations, the communication region of each anchor point can be treated as a square, as illustrated in Figure 5b, and the intersection region of the incoming neighbor can then be calculated.

Step 2. Calculating the distance from the unknown node to each anchor node

In this step, the idea of the DV-Hop localization algorithm is adopted, and the product of hop count and hop distance between nodes is used as the distance from an unknown node to an anchor node. The hop distance formula is shown in Equation (10), and the distance from the unknown node to the anchor node is calculated in Equation (11).

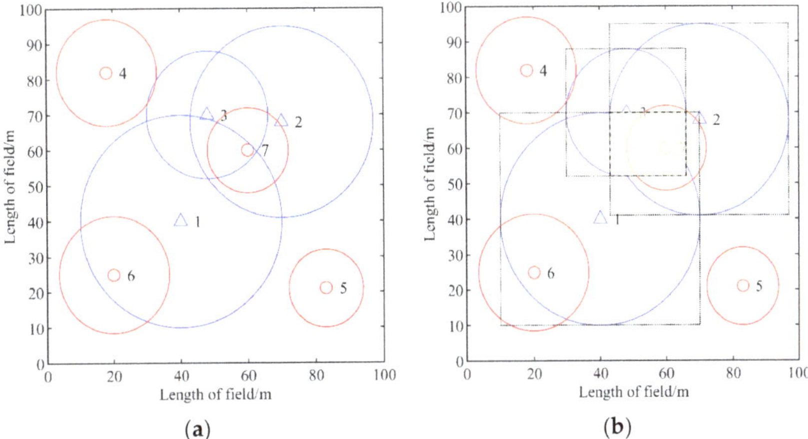

Figure 5. Impact of heterogeneity of communication radius on network communication: (**a**) original communication radius; (**b**) square communication radius.

$$Hopsize_i = \frac{\sum_{i \neq k} \sqrt{(x_i - x_k)^2 + (y_i - y_k)^2}}{\sum_{i \neq k} h_{ik}} \tag{10}$$

$$d_{ij} = Hopsize_{ij} \cdot h_{ij} \tag{11}$$

where (x_i, y_i), (x_k, y_k) are the coordinates of anchors I and k, respectively, and h_{ik} represents the number of hops between anchors I and k. $Hopsize_i$ is the hop distance of anchor I, $Hopsize_{ij}$ denotes the hop distance of anchor I with the least number of hops from unknown node j, and d_{ij} is the distance between unknown node j to anchor i.

Step 3. Calculating coordinates of unknown node

To begin with, the fitness function is defined as in Equation (12). Then, the swarm intelligence optimization algorithm is used to find the unknown node coordinates, with the coordinates with the lowest fitness values being selected as the unknown node coordinates. If the distance value in Step 2 is guaranteed to be constant, the higher the performance of the chosen swarm intelligence optimization algorithm, the more accurate the positioning.

$$fitness_j = \sum_{i=1}^{Na} \left| \sqrt{(\tilde{x}_j - x_i)^2 + (\tilde{y}_j - y_i)^2} - d_{ij} \right| \tag{12}$$

where j represents the unknown node numbers, N_a is the total number of anchors, and (x_j, y_j) represents the coordinates of unknown nodes randomly generated.

4.4. Performance Evaluation

To verify the superiority of ISSA for node localization in HWSNs, we compared it with LS and 15 other meta-heuristic algorithms, including HHO [13], GWO [14], BOA [15], SSA [16], Gold-SA [23], PSO [24], GTO [26], sine cosine algorithm (SCA) [31], DE [32], Archimedes optimization algorithm (AOA) [33], whale optimization algorithm (WOA) [34], elephant herding optimization (EHO) [35], marine predators algorithm (MPA) [36], tunicate swarm algorithm (TSA) [37], and Coot optimization algorithm (COOT) [38]. All meta-heuristic algorithms have a population size of 30, an iteration number of 50, and other parameters that are consistent with their original literature. The area of HWSNs is $100 \times 100 \ m^2$, the number of anchor nodes is 25, and the node communication radius is a random value within [15,29], and the normalized root mean square error (NRMSE) is used to evaluate the positioning performance as shown in Equation (13).

$$NRMSE = \frac{1}{Nu}\sum_{j=1}^{Nu}\frac{1}{r_j}\sqrt{(x_j - \tilde{x}_j)^2 + (y_j - \tilde{y}_j)^2}$$

(13)

where Nu represents the number of unknown nodes and (x_i, y_i), and $(\tilde{x}_j, \tilde{y}_j)$ are the true and estimated coordinates of the unknown node, respectively.

The NRMSE and the time required to compute the coordinates of unknown nodes using LS or different meta-heuristic algorithms when all nodes in HWSNs have the same coordinate positions are shown in Table 4. As far as the NRMSE is concerned, the average localization error of LS is higher than that of all other metaheuristic algorithms, reaching 55.57%. This indicates that using meta-heuristic algorithms to solve the coordinates of unknown nodes can indeed reduce the localization error of the nodes. The localization error obtained using different metaheuristic algorithms also varies. The best-performing ISSA has a 13.19% reduction in the average positioning error compared with the worst performing EHO. The top four average positioning errors are ISSA, GTO, SSA, and DE in order, and their average positioning errors are 41.38%, 41.51%, 41.68%, and 41.78%, respectively. Although the NEMSE of the top four meta-heuristic algorithms is not very different, there is a big difference in the positioning time required by each of them.

Table 4. Positioning results of 17 algorithms.

	NRMSE			Time		
	AVG	**STD**	**Rank**	**AVG**	**STD**	**Rank**
LS	0.5557	0.0945	17	0.0903	0.0358	1
PSO	0.4545	0.0592	15	1.0545	0.2449	5
DE	0.4178	0.0610	4	1.4722	0.1060	13
SCA	0.4252	0.0602	6	1.1997	0.1844	9
Gold-SA	0.4245	0.0633	5	1.1809	0.2874	8
AOA	0.4408	0.0606	8	0.9484	0.3666	3
GWO	0.4414	0.0572	12	1.1724	0.1841	7
WOA	0.4448	0.0574	13	1.1524	0.1596	6
EHO	0.5457	0.0484	16	1.3273	0.1379	10
BOA	0.4389	0.0542	7	1.4387	0.0863	11
SSA	0.4168	0.0596	3	0.9542	0.1997	4
MPA	0.4409	0.0595	10	3.1366	0.4165	16
TSA	0.4414	0.0572	11	0.7694	0.0841	2
COOT	0.4409	0.0569	9	1.4474	0.4420	12
HHO	0.4453	0.0548	14	2.3375	0.3184	15
GTO	0.4151	0.0611	2	10.3031	0.4517	17
ISSA	0.4138	0.0590	1	1.6637	0.4558	14

In terms of search time, LS takes the shortest time of 0.04 s, while GTO takes the longest time of 10.3 s. The times required for ISSA, GTO, SSA, and DE localization are 1.66 s, 10.3 s, 0.95 s, and 1.47 s, respectively. It can be seen that, when solving the node coordinates of HWSNs, it is difficult to achieve both accuracy and time optimality. However, the localization accuracy is the most critical index in the node localization of HWSNs. Within a certain range, the search time requirement can be reduced appropriately. In summary, the ISSA proposed in this paper is more suitable for node localization in HWSNs.

4.5. Effect of Parameter Variation on Localization Accuracy in HWSNs

Figure 6 depicts the effect on NRMSE when the number of nodes and the ratio of anchor points are varied in a deployment area of 100×100 m^2. The 10–30 in Figure 6a,b indicates that the communication radius of the nodes is a random value within [10, 30], and 30–30 indicates that the communication radius of the nodes is all 30 m. The ratio of anchor points in Figure 6a is 25, and the total number of nodes in Figure 6b is 100. ISSA is used to calculate the coordinates of unknown nodes with 50 iterations, and the number of individuals in the population is 30.

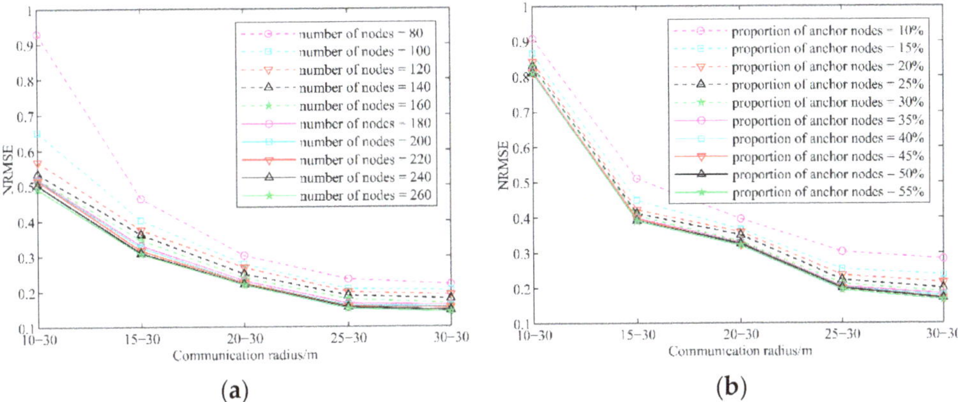

Figure 6. Effect of parameter variation on positioning accuracy: (**a**) change in number of nodes; (**b**) change of anchor node ratio.

From Figure 6, the NRMSE decreases as the heterogeneous range of node communication radius decreases, regardless of the number of nodes and the ratio of anchor points taken. From Figure 6a, the NRMSE decreases continuously when the number of nodes increases sequentially, but the magnitude of its decrease also becomes smaller. When the number of nodes increases to 160, the average variation of NRMSE on different communication intervals is less than 1%. From Figure 6b, the NRMSE also decreases as the proportion of anchor points increases, but the magnitude of its decrease also becomes slower. When the anchor point ratio is 25%, the reduction of NRMSE tends toward 0. Therefore, the optimal number of nodes is 160, and the optimal anchor point ratio is 25% when using the localization algorithm described in this section to solve the localization problem.

5. Conclusions

In this paper, for the node localization problem in HWSNs, we compared the localization accuracy and localization time consumption of 15 common meta-heuristic algorithms and found that the comprehensive performance of SSA is the best. Thus, a multi-strategy improved sparrow search algorithm is proposed to address the problems of SSA. To improve the global exploration capability of SSA, the golden sine strategy was introduced into the producer's position update method. Meanwhile, to accelerate the convergence speed of SSA, the idea of individual optimality of particle swarm was introduced to the location update method of the investigator. To prevent SSA from falling into local optimality, Gaussian perturbation was applied to the globally optimal sparrow individuals. A total of 23 benchmark functions and 5 meta-heuristics were chosen to evaluate ISSA's optimization performance. In terms of the average value and standard deviation of the search results, ISSA achieved first place or was tied for first place on a total of 20 benchmark functions. Except for F14, the convergence speed of ISSA is not weaker than the rest of the comparison algorithms. In addition, the Wilcoxon rank sum test and the average value were combined in order to evaluate the performance of the algorithms. From a statistical point of view, the number of benchmark functions in which ISSA outperforms PSO, Gold-SA, HHO, GTO, and SSA is 14, 10, 13, 6, and 13, respectively.

Finally, ISSA was used to solve the problem of solving unknown node coordinates for HWSNs coordinate calculation. Simulation experiments showed that the localization accuracy obtained using ISSA to compute node coordinates is the best among the 15 metaheuristic algorithms and LS, and it reduces the localization error by 14.19% compared with LS. Changing the internal parameters of HWSNs, it can be found that increasing the number of nodes and increasing the proportion of anchor nodes can improve the accuracy of the localization algorithm. However, when the number of nodes reaches 160 and the proportion of anchor nodes reaches 25%, the enhancement effect will be significantly weakened.

Unfortunately, applying the proposed ISSA to the node localization of HWSNs improves the localization accuracy but also increases the time required for localization. In our future work, we will further optimize the search mechanism of ISSA to reduce the search time during node localization. In addition, we will also focus on the distance estimation of the first stage of the HWSNs localization problem to further improve the localization accuracy.

Author Contributions: Conceptualization, H.Z. and J.Y.; formal analysis, J.Y.; investigation, W.W. and Y.F.; methodology, H.Z.; software, T.Q.; supervision, W.W.; validation, H.Z., W.W. and Y.F.; writing—original draft, H.Z.; writing—review and editing, H.Z., J.Y. and Z.L. All authors have read and agreed to the published version of the manuscript.

Funding: This work was supported in part by the NNSF of China (No. 61640014, No. 61963009), the Industrial Project of Guizhou province (No. Qiankehe Zhicheng [2022]Yiban017, [2019] 2152), the Innovation Group of the Guizhou Education Department (No. Qianjiaohe KY [2021]012), the Science and Technology Fund of Guizhou Province (No. Qiankehejichu [2020]1Y266), the CASE Library of IOT (KCALK201708), and the IOT platform of the Guiyang National High technology industry development zone (No. 2015).

Institutional Review Board Statement: Not applicable.

Informed Consent Statement: Not applicable.

Data Availability Statement: Data are contained within the article. The data presented in this study can be requested from the authors.

Conflicts of Interest: The authors declare no conflict of interest.

References

1. Phoemphon, S.; So-In, C.; Leelathakul, N. Improved distance estimation with node selection localization and particle swarm optimization for obstacle-aware wireless sensor networks. *Expert Syst. Appl.* **2021**, *175*, 114773. [CrossRef]
2. Nemer, I.; Sheltami, T.; Shakshuki, E.; Elkhail, A.A.; Adam, M. Performance evaluation of range-free localization algorithms for wireless sensor networks. *Pers. Ubiquit. Comput.* **2021**, *25*, 177–203. [CrossRef]
3. Nithya, B.; Jeyachidra, J. Low-cost localization technique for heterogeneous wireless sensor networks. *Int. J. Commun. Syst.* **2021**, *34*, e4733. [CrossRef]
4. Tu, Q.; Liu, Y.; Han, F.; Liu, X.; Xie, Y. Range-free localization using Reliable Anchor Pair Selection and Quantum-behaved Salp Swarm Algorithm for anisotropic Wireless Sensor Networks. *Ad Hoc Netw.* **2021**, *113*, 102406. [CrossRef]
5. Liu, Z.K.; Liu., Z. Node self-localization algorithm for wireless sensor networks based on modified particle swarm optimization. In Proceedings of the 27th Chinese Control and Decision Conference (2015 CCDC), Qingdao, China, 23–25 May 2015; pp. 5968–5971.
6. Chai, Q.; Chu, S.; Pan, J.; Hu, P.; Zheng, W. A parallel WOA with two communication strategies applied in DV-Hop localization method. *EURASIP J. Wirel. Commun.* **2020**, *2020*, 50. [CrossRef]
7. Cui, L.; Xu, C.; Li, G.; Ming, Z.; Feng, Y.; Lu, N. A high accurate localization algorithm with DV-Hop and differential evolution for wireless sensor network. *Appl. Soft Comput.* **2018**, *68*, 39–52. [CrossRef]
8. Assaf, A.E.; Zaidi, S.; Affes, S.; Kandil, N. Low-Cost Localization for Multihop Heterogeneous Wireless Sensor Networks. *IEEE Trans. Wirel. Commun.* **2016**, *15*, 472–484. [CrossRef]
9. Wu, W.; Wen, X.; Xu, H.; Yuan, L.; Meng, Q. Efficient range-free localization using elliptical distance correction in heterogeneous wireless sensor networks. *Int. J. Distrib. Sens. Netw.* **2018**, *14*, 1–9. [CrossRef]
10. Bhat, S.J.; Venkata, S.K. An optimization based localization with area minimization for heterogeneous wireless sensor networks in anisotropic fields. *Comput. Netw.* **2020**, *179*, 107371. [CrossRef]
11. Goel, L. An extensive review of computational intelligence-based optimization algorithms: Trends and applications. *Soft Comput.* **2020**, *24*, 16519–16549. [CrossRef]
12. Zhang, J.; Wang, J.S. Improved Whale Optimization Algorithm Based on Nonlinear Adaptive Weight and Golden Sine Operator. *IEEE Access* **2020**, *8*, 77013–77048. [CrossRef]
13. Heidari, A.A.; Mirjalili, S.; Faris, H.; Aljarah, I.; Mafarja, M.; Chen, H. Harris hawks optimization: Algorithm and applications. *Future Gener. Comput. Syst.* **2019**, *97*, 849–872. [CrossRef]

14. Mirjalili, S.; Mirjalili, S.M.; Lewis, A. Grey Wolf Optimizer. *Adv. Eng. Softw.* **2014**, *69*, 46–61. [CrossRef]
15. Arora, S.; Singh, S. Butterfly optimization algorithm: A novel approach for global optimization. *Soft Comput.* **2018**, *23*, 715–734. [CrossRef]
16. Xue, J.; Shen, B. A novel swarm intelligence optimization approach: Sparrow search algorithm. *Syst. Sci. Control. Eng.* **2020**, *8*, 22–34. [CrossRef]
17. Liu, G.; Shu, C.; Liang, Z.; Peng, B.; Cheng, L. A Modified Sparrow Search Algorithm with Application in 3d Route Planning for UAV. *Sensors* **2021**, *21*, 1224. [CrossRef]
18. Zhou, J.; Chen, D. Carbon Price Forecasting Based on Improved CEEMDAN and Extreme Learning Machine Optimized by Sparrow Search Algorithm. *Sustainability* **2021**, *13*, 4896. [CrossRef]
19. An, G.; Jiang, Z.; Chen, L.; Cao, X.; Li, Z.; Zhao, Y.; Sun, H. Ultra Short-Term Wind Power Forecasting Based on Sparrow Search Algorithm Optimization Deep Extreme Learning Machine. *Sustainability* **2021**, *13*, 10453. [CrossRef]
20. Yang, X.; Liu, J.; Liu, Y.; Xu, P.; Yu, L.; Zhu, L.; Chen, H.; Deng, W. A Novel Adaptive Sparrow Search Algorithm Based on Chaotic Mapping and T-Distribution Mutation. *Appl. Sci.* **2021**, *11*, 11192. [CrossRef]
21. Yuan, J.; Zhao, Z.; Liu, Y.; He, B.; Wang, L.; Xie, B.; Gao, Y. DMPPT Control of Photovoltaic Microgrid Based on Improved Sparrow Search Algorithm. *IEEE Access* **2021**, *9*, 16623–16629. [CrossRef]
22. Liu, J.; Wang, Z. A Hybrid Sparrow Search Algorithm Based on Constructing Similarity. *IEEE Access* **2021**, *9*, 117581–117595.
23. Tanyildizi, E.; Demir, G. Golden Sine Algorithm: A Novel Math-Inspired Algorithm. *Adv. Electr. Comput. Eng.* **2017**, *17*, 71–78. [CrossRef]
24. Kennedy, J.; Eberhart, R. Particle swarm optimization. In Proceedings of the ICNN'95—International Conference on Neural Networks, Perth, WA, Australia, 27 November–1 December 1995; pp. 1942–1948.
25. Xiao, S.; Wang, H.; Wang, W.; Huang, Z.; Zhou, X.; Xu, M. Artificial bee colony algorithm based on adaptive neighborhood search and Gaussian perturbation. *Appl. Soft Comput.* **2021**, *100*, 106955. [CrossRef]
26. Abdollahzadeh, B.; Soleimanian Gharehchopogh, F.; Mirjalili, S. Artificial gorilla troops optimizer: A new nature-inspired metaheuristic algorithm for global optimization problems. *Int. J. Intell. Syst.* **2021**, *36*, 5887–5958. [CrossRef]
27. Zhang, M.; Wang, D.; Yang, J. Hybrid-Flash Butterfly Optimization Algorithm with Logistic Mapping for Solving the Engineering Constrained Optimization Problems. *Entropy* **2022**, *24*, 525. [CrossRef]
28. Jamil, M.; Yang, X. A literature survey of benchmark functions for global optimisation problems. *Int. J. Math. Model. Numer. Optim.* **2013**, *4*, 150–194. [CrossRef]
29. Rashedi, E.; Nezamabadi-Pour, H.; Saryazdi, S. GSA: A Gravitational Search Algorithm. *Inf. Sci.* **2009**, *179*, 2232–2248. [CrossRef]
30. García, S.; Molina, D.; Lozano, M.; Herrera, F. A study on the use of non-parametric tests for analyzing the evolutionary algorithms' behaviour: A case study on the CEC'2005 Special Session on Real Parameter Optimization. *J. Heuristics* **2009**, *15*, 617–644. [CrossRef]
31. Mirjalili, S. SCA: A Sine Cosine Algorithm for solving optimization problems. *Knowl.-Based Syst.* **2016**, *96*, 120–133. [CrossRef]
32. Storn, R.; Price, K. Differential Evolution—A Simple and Efficient Heuristic for global Optimization over Continuous Spaces. *J. Glob. Optim.* **1997**, *11*, 341–359. [CrossRef]
33. Hashim, F.A.; Hussain, K.; Houssein, E.H.; Mabrouk, M.S.; Al-Atabany, W. Archimedes optimization algorithm: A new metaheuristic algorithm for solving optimization problems. *Appl. Intell.* **2020**, *51*, 1531–1551. [CrossRef]
34. Mirjalili, S.; Lewis, A. The Whale Optimization Algorithm. *Adv. Eng. Softw.* **2016**, *95*, 51–67. [CrossRef]
35. Wang, G.-G.; Deb, S.; Coelho, L.d.S. Elephant Herding Optimization. In Proceedings of the 2015 3rd International Symposium on Computational and Business Intelligence (ISCBI), Bali, Indonesia, 7–9 December 2015; pp. 1–5.
36. Faramarzi, A.; Heidarinejad, M.; Mirjalili, S.; Gandomi, A.H. Marine Predators Algorithm: A nature-inspired metaheuristic. *Expert Syst. Appl.* **2020**, *152*, 113377. [CrossRef]
37. Kaur, S.; Awasthi, L.K.; Sangal, A.L.; Dhiman, G. Tunicate Swarm Algorithm: A new bio-inspired based metaheuristic paradigm for global optimization. *Eng. Appl. Artif. Intell.* **2020**, *90*, 103541. [CrossRef]
38. Naruei, I.; Keynia, F. A new optimization method based on COOT bird natural life model. *Expert Syst. Appl.* **2021**, *183*, 115352. [CrossRef]

 *applied sciences*

Article

Identification of Opinion Leaders and Followers—A Case Study of Green Energy and Low Carbons

Chun-Che Huang [1], Wen-Yau Liang [2,*], Po-An Chen [1] and Yi-Chin Chan [3]

[1] Department of Information Management, National Chi Nan University, Pu-Li 54561, Taiwan; cchuang@ncnu.edu.tw (C.-C.H.); annieab.chen@gmail.com (P.-A.C.)

[2] Department of Information Management, National Changhua University of Education, Changhua 50074, Taiwan

[3] Program in Strategy and Development of Emerging Industries, National Chi Nan University, Pu-Li 54561, Taiwan; s106245903@mail1.ncnu.edu.tw

* Correspondence: wyliang@cc.ncue.edu.tw

Received: 21 October 2020; Accepted: 24 November 2020; Published: 26 November 2020

Abstract: In recent years, with the development of Web2.0, enterprises, government agencies, and traditional news media, which have been positively influenced by opinion leaders, have been dedicated to understanding leaders' opinions on the web in order to seek convergence. Specifically, with the increase of environmental awareness, the introduction of green energy and carbon reduction technology has become an important issue. Consequently, studies identifying opinion leaders and followers who are interested in green energy and low carbon have become important. This study aims to find a solution that can identify the characteristics of opinion leaders and followers that can be widely used, which will help certain public policies or issues to be more effectively disseminated in the future. To model the characteristics of opinion leaders and their influence on followers, this study uses a dual matrix. The interaction patterns are recognized among opinion leaders and followers, with the aim of developing public policy to promote green energy and low carbon emissions. A case is studied to validate the superiority of the proposed solution approach. With the proposed approach, a (business) organization can identify and access opinion leaders and their followers. Through communication, these organizations can absorb strain and preserve functions despite the presence of adversity. This study also clearly demonstrates its contribution and novelty through comparisons with the existing alternative method.

Keywords: green energy and low carbon; opinion leaders; followers; social media; matrix method; intelligent systems

---

1. Introduction

With the rise of communication technology, people are utilizing platforms such as content sharing sites, blogs, social networking, and wikis to create, modify, share, and discuss Internet content. Social media provides flexible platforms that play key roles in energizing collective action in movements [1]. This represents the social media phenomenon, which can significantly impact society and industry, e.g., firms' reputations, sales, and even survival [2]. Within the discussions on social media, certain individuals influence others and thus emerge as opinion leaders. Opinion leaders have great impacts and influence on social media. Organizations can take advantage of these predispositions through marketing research and public relations, nurturing opinion leaders or advocates, placing and creating advertisements, developing new products and lowering the cost-to-serve [3]. On the Internet, the power of these leaders is increasing larger and sequentially influencing entire societies through calls to protest, promotion of policy and decision-making, which was defined as the fifth right in the "Towards a Civil Society" seminar [4].

Appl. Sci. **2020**, *10*, 8416; doi:10.3390/app10238416   www.mdpi.com/journal/applsci

The world must confront the energy crisis and air pollution. Discussions about energy issues are increasing. These discussions range from nuclear energy, thermal power, hydropower, and other forms of green energy and low carbon technology, including wind, solar, tidal, and biomass geothermal energy issues. In Taiwan, whenever an energy crisis occurs, energy charges increase. Anti-nuclear positions and other energy issues are discussed broadly. Therefore, the Taiwan government tries to understand people's needs and questions.

On the Internet, the roles of opinion leaders and followers in the formation of these issues cannot be neglected. According to the "theory of two-step flow" [5] and Rosen's definition of opinion leaders' characteristics, "social media initially pass the information to opinion leaders, then opinion leaders spread the information to followers and influence their attitudes" [6]. Thus, when followers follow opinion leaders, the formers' judgments and attitudes will be influenced and changed by opinion leaders. This study defines opinion leaders as people or social media with high social status who are able to influence followers. This study defines followers as the users who follow certain issues, publish related discussions and add their own ideas. They spread, repost or blindly follow the behaviors of opinion leaders.

Most previous research of opinion leaders focuses on the commercial domain rather than on nonprofit-related policies such as energy policy [7]. In the promotion of many public policies through online postings, it is difficult to clearly identify opinion leaders and followers, which greatly reduces the effectiveness of communication. Based on the community attributes of opinion leaders and whether they can successfully resonate, this study aims at providing a method to try to identify who are opinion leaders or who are likely to become opinion leaders in social media, and who are followers. Relational matrix analysis is used to represent the relationship between opinion leaders and followers in social media and to identify the collection of opinion leaders and potential opinion leaders.

Furthermore, previous studies [8] have used quantitative methods of analysis. One example is the SuperedgeRank algorithm. However, this algorithm not only has difficulty identifying potential opinion leaders effectively but also neglects how opinion leaders influence followers and how relationships between opinion leaders and followers are characterized. It also ignores the increasingly important role played by intelligent systems such as algorithms.

Although the literature on green energy is rapidly increasing, many studies suggest that this problem needs to be dealt with by considering a broader perspective [9]. This study not only examines the issue from the perspective of intelligent systems such as algorithms but also identifies the roles of opinion leaders and followers on social media in relation to the introduction of green energy and carbon reduction technology, with the aim of developing public policy to promote green energy and low carbon emissions. This study is novel not only because it takes quantitative factors and tradition clustering approaches into account but because it also analyzes posts, poster characteristics and their interactive relationships on social media. This study reviews relevant literature in Section 2. In Section 3, we propose a method to identify opinion leaders and their followers based on their interactions on social networks. The interaction patterns are also identified. An energy case is studied in Section 4 to validate the proposed solution approach and enhance communication effectiveness between government policymakers and people's desires. The discussion is summarized in Section 5, and Section 6 concludes this study.

This research contributes to finding a solution to easily identify the characteristics of opinion leaders and followers in the case of online posts related to green energy and low-carbon policies. Once certain public policies need to be effectively disseminated, they can be widely used. Using the same model and the solution approach, the results of this study can be extended from the green energy low carbon issue to other social issues. Furthermore, this study provides a new perspective to deal with the effective identification of opinion leaders and followers, at the same time, promotes the "theory of two-step flow" to add another research perspective in the academic field.

2. Literature Review

2.1. Green Energy and Low Carbon

Global warming, unexpected climate change, dwindling energy resources and unprecedented amounts of air pollution have become critical problems. The United Nations' 2030 Sustainable Development Goals show that a sustainable modern electricity grid [10], reduction of CO_2 emissions [11] and the carbon footprint of human mobility to a sustainable level [12] etc., are key parts. In addition, exhaustion of fossil fuel is viewed as a big challenge of human development [13] since energy is an expensive resource that is becoming more scarce with increasing population and demand [14]. Green energy could help governments reduce the dependency on energy importation, improve the variety of production resources and advance sustainable environmental development. Moreover, the usage of rich green energy could benefit economies significantly [15].

In the past, studies of green energy and low carbon have focused on issues of energy itself and energy systems, for example, integration of energy systems [16], reliability of the power distribution system [17], uptake of biomass energy [18], and so on. Few studies have focused on social opinions about green energy and low carbon. In recent years, due to awareness of the environment, the public has started to care more about the environment and quality of life [19]. Social media has strengthened community among people and emerged as a platform to spread messages quickly and powerfully. In the era of Web 2.0, massive public opinion is increasingly generated on the Internet [20]. Therefore, the study of the role of social media in green energy and low carbon social issues is very important.

2.2. Opinion Leaders

In the era of the Internet, opinion leaders enhance content sharing. In fact, almost all of the content is generated by opinion leaders (the 90–91% law) [21]. What makes opinion leaders so important on social networks is their ability to informally influence others' attitudes and behaviors [22–25]. Opinion leaders usually have access to far more information on a certain topic and have professional experience with the topic. Rosen defined the characteristics of opinion leaders proposed the acronym ACTIVE. ACTIVE stands for the six characters of opinion leaders: ahead in adoption, connected, travelers, information-hungry, vocal, and exposed to the media [6].

In a recent qualitative survey carried out through focus groups, Katz and Lazarsfeld proposed the "theory of two-step flow" in 1995 and pointed out that opinion leaders are situated between social media and the majority of people. The information first reaches the opinion leaders or influencers [26], who then introduce it to the wider population [5]. Followers are those who are affected and change their behaviors and attitudes when receiving the information [27]. The followers are enormously influenced by opinion leaders in terms of changing their attitudes and behaviors [25].

Based on the above study, we elaborate on the attributes of opinion leaders as follows: Their life experience and understanding of knowledge are rich and thorough and a majority of them are highly educated. Moreover, they have strong social skills, strong connections with the broad masses, and good reputations due to their professionalism and knowledge. They have great influence and appealing power. They exhibit sensitivity to information, willingness to accept new things and an innovative spirit.

2.3. Opinion Leaders Identification

Many theories have been put forward about social networks, but few address the issue of opinion leader identification [28]. According to a previous literature review, opinion leaders are simply determined based on some visible user activities, and other factors that allow a user to become an opinion leader are ignored [28,29]. Studies of Internet opinion leaders have also mainly focused on the role of Internet opinion leaders in spreading the news and in the Internet world of word mouth marketing [23]. Consensus has not yet been reached in the analysis of Internet opinion leaders. Few efforts have been taken to create a computer-based model to identify and analyze the opinion leaders in an Internet community, and the studies that have been undertaken on this issue have failed to reach an in-depth level [30]. At present, studies related to the TwitterRank algorithm based on PageRank [29] and the contribution of information to InfluenceRank [31] and weighted Page-Rank [32] make use of the network construction of user interaction, but they neglect the users' inherent features [33].

In addition, from the perspective of data mining, the identification of opinion leaders is a cluster problem. However, the aforementioned studies consider the relationships of people to be a social network. Engagement is used as an effective degree to measure user interaction with an organization. Basic interactions include commenting on contents, sharing contents, or "liking" or "favoriting" content. A core KPI for social media is that engagement is high, as this would indicate that organizations are producing content that users find interesting enough to spend additional time on [34]. Unfortunately, when applying the cluster problem to social networks, previous studies have only taken quantitative factors, and traditional clustering approaches into account, e.g., support vector machines [35], k-means [36], partitioning around medoids [37], fuzzy c-means [38], and so on have been used to resolve quantitative clusters. However, qualitative characteristics are not yet considered, and only static data have been analyzed. To study the qualitative characteristics of opinion leaders and the impact of opinion leaders on followers, [39] evaluates whether every speaker in social media satisfies the characteristics of an opinion leader. By observing the relational matrix, the interacting relations between users in social media are analyzed, and opinion leaders and followers are identified. However, there are no theoretical background axioms implied in [39], specifically from the perspective of communication to validate the results.

2.4. Opinion Leaders Identification Algorithms

Ma and Liu [40] used the SuperedgeRank algorithm to analyze the attributes of three seed networks and identify opinion leaders on the Fukushima nuclear issue. In another study, Jiang et al. [41] designed and implemented a BBS opinion leader mining system based on an improved PageRank algorithm using MapReduce. Ziyi et al. [30] adopted the core algorithm of the Internet searching–PageRank model and, by combining the analysis of the influence of linguistic data and sentimental preference, put forward a method to identify Internet opinion leaders; they also verified the method by carrying out an empirical study. Cheng et al. [42] combined influence with sentimental analysis based on the content of posts and filtered opinion leaders by combining the PR values of the PageRank algorithm and recognition degree, abbreviating the IS Rank algorithm. Deng et al. [43] constructed a SINA Micro Blog APIs based Micro Blog crawling and analysis tool, and a node betweenness approximation computation method was proposed, offering better accuracy and less running time to detect core opinion leaders on Micro Blog graphs.

PageRank is an excellent sorting algorithm, but its running speed decreases significantly with the increase of the number of data nodes. Jing and Lizhen [33] proposed a hybrid data mining approach based on user features and interaction networks, which includes three parts: a way to analyze users' authority, activity and influence, a way to consider the orientation of sentiment in an interaction network and a combined method based on the HITS algorithm for identifying microblog opinion leaders [33]. Chu et al. [44] researched social networks to access the influence of tobacco opinion leaders on followers and found that followers are a vulnerable group. They are young and low educated. Followers are easily influenced by opinion leaders. Therefore, anti-smoking education to stay away from tobacco can educate them on social media. Obviously, opinion leaders on the Internet have considerable influence on followers, and opinion leaders are often used in marketing in the e-commerce industry. The research of Lin et al. [45] found that opinion leaders can use their influence to act as important promoters of products and services. It is recommended that companies or corporate managers choose to cooperate with opinion leaders of a certain type of forum to promote products or services. What is the impact of the levels of followers' trust in opinion leaders on the resulting influence? Zhao et al. [46] used opinion dynamics theory to study the influence of trust in opinion leaders. His research found that followers' trust in opinion leaders determines opinion leader influence. It is suggested that if the communication effect of e-commerce is pursued, the key premise can increase the trust of opinion leaders.

Not only have many studies investigated opinion leaders in the e-commerce field, but the role and function of opinion leaders have attracted attention in politics and the public domain. Aleahmad et al. [43] examined the political field by proposing the effective OLfinder algorithm. The researchers found that this algorithm can not only find opinion leaders in social networks but also calculate their popularity. Many people are curious about why opinion leaders like to play the role of opinion leaders. Winter et al. [47] examined people who disseminate opinions about politics or public affairs on the Internet and identified these people as opinion leaders who try to influence the psychological motivations and personality characteristics of followers. This study found that opinion leaders have strong psychological motivations to actively express themselves and persuade others, making them like to play the role of opinion leaders. In addition, in social network analysis, centrality methods have been applied to measure the importance of nodes in a network whereby nodes with higher centrality can influence others more significantly [48,49].

All in all, past relevant research either used a certain social measurement method based on interview self-reports or questionnaire surveys or used quantitative clustering techniques to identify opinion leaders. Few studies have actually investigated online posts to identify opinion leaders and the social patterns of interaction between opinion leaders and followers. Thus, it is important to propose a method to analyze posts, posters' characteristics and their interactive relationships in social media.

3. Methodology

3.1. Modeling Interaction between Opinion Leaders and Followers

Between users, matrix M is a relational matrix.

$$M = [T|I] \tag{1}$$

We set the row index as i and the column index as j in the matrix. The n users in the set are represented by C = {1 … n}. In the matrix T, the elements are composed of counts of responses and being responded to. Additionally, between users and social community support level, the elements in the matrix I are the influence factors

Matrix T shows us the counts of both responses and being responded to between n users. T_ij refers to counts of responses of $user_i\_i$ to $user_j\_j$ where $i \neq j$. When T_ij refers to counts of total posts of $user\_i$ where $i = j$. i refers to the index of the users who responds to other user's opinions, and j refers to the index of the users whose opinions are responded to.

Matrix I indicates the power of users to influence and be influenced. I_ij refers to the influence of $user\_i$ on $user\_j$ where $i \neq j$, i refers to the index of a user who influences other people, and j refers to the index of a user who is influenced by others. The influence power can be classified into three different patterns, including job position, professional knowledge and social community support level [39]. Each criterion has three levels, no influence (NI), general influence (GI), and high influence (HI) (Table 1). For a higher job title with more professional knowledge and a high social community support level, we define influence power as high influence (HI). For a general job title with popular professional knowledge and neutral social community support level, the overall influence power is general (GI). If the user has no job and inaccurate professional knowledge or social community support, we define that the influence level as having no influence (NI). However, some users have privacy settings or discuss issues anonymously, so our study could not gather their background information. Fortunately, the proposed approach can serve as a flexible model with the missing part left blank.

Table 1. Influence power table (NI: No Influence, GI: General Influence, HI: High Influence.).

			Social Status			
		−	General	+		
Professional knowledge	Wrong knowledge	NI	NI	NI	Low social community support	Social community Support
	Popular knowledge	NI	NI	GI	Low social community support	
	Professional knowledge	NI	GI	GI	Low social community support	
Professional knowledge	Wrong knowledge	NI	NI	GI	Neutral social community support	Social community Support
	Popular knowledge	NI	GI	GI	Neutral social community support	
	Professional knowledge	GI	GI	HI	Neutral social community support	
Professional knowledge	Wrong knowledge	NI	GI	HI	High social community support	Social community Support
	Popular knowledge	GI	HI	HI	High social community support	
	Professional knowledge	HI	HI	HI	High social community support	

In matrix I, I_ij refers to a user's social community support level where $i = j$. In this study, social community support is divided into three levels. These three levels are relatively well-known, and well-followed social media that receives good attention are which are classified as having high social community support (H). Less-followed, less-known social media with low attention is classified as low (L). However, we were unable to gather users' social community support levels because of some users' anonymous discussions or privacy settings. Our study defines the social community support level of these users as missing (O).

A non–follower, a user with a negative speech count, is defined in this study. Meanwhile, when contents are responded to negatively, the user is listed as a non-opinion leader. The relationship between opinion leaders and followers refers to a mutual relationship between users. However, one user may not respond to or express ideas to others' speech content on social media. One opinion leader may not be an opinion leader of all users. Therefore, if there is no relationship between users, we cannot distinguish whether they are opinion leaders or followers. In this study, the groups of opinion leaders and followers are be judged base on interaction. The users with mutual influence are classified as one group to analyze whether there is an opinion leader and a follower in the group. If there is no mutual influence, no group is formed.

3.2. Opinion Leaders and Followers' Social Patterns

Six axioms are proposed to classify opinion leaders, influencers, followers and interaction patterns. Notations are presented in Table 2.

Table 2. Notations table.

Notations:
N_{rsp} : The total number of responses in matrix T.
$\overline{N}_{rsp}$: The average total number of responses in matrix T.
N_g : The total number of members that respond in group g.
$\overline{N}_g$: The average total number of members' responses in group g.
$T(US_i)$: The number of $user'_i s$ posts, the total counts of $column_i$ in matrix g.
$T(UR_i)$: The number of $user'_i s$ responses, the total counts of row_i in matrix T.

Table 2. *Cont.*

Notations:

USI_{ij} : The influence of $user'_i$s speech to $user_j i$, where i is the index of the column, and j is the index of the row.

URI_{ij} : The influence of $user'_i$s responses from $user_j$, where j is the index of the column and I is the index of the row.

$SI_i\{INFLUENCE\}$: The number of $user'_i$s influences (NI, GI, HI).

SI_{ol} : Number of the opinion leaders' posts in matrix I.

SI_f : Number of followers' posts in matrix I.

$SI_{ol}\{influence\ power\}$: The number of opinion leaders' influences (NI, GI, HI).

$SI_f\{NI\}$: The number of opinion leaders' influences (NI, GI, HI).

UNI_i: The number of $user'_i$s posts with NI influence power and responses in matrix T are zero.

The axioms formulated by the systematic rolling analysis of big data over the years. Experts determine the criteria and select thresholds for them, for example, the positive or negative degree of the counts of responses to and responded to. The following axioms are proposed to recognize opinion leaders, influencers, and followers.

Axiom1—Opinion leader.

If $\overline{N}_{rsp} < UR_i$ and $UR_i > 2 \times \overline{N}_g$ and $UNI < 1/2 \times N_g$, then the $user_i$ is an opinion leader.

To be an opinion leader, $user_i$'s (UR_i) response total must be higher than the average total response ($\overline{N}_{rsp}$), and $user_i$ must have more than twice the number of responses in the group ($\overline{N}_g$). Moreover, $user'_i$s NI speech and the count in matrix $T(US_i)$ less than 1 cannot be more than half of the total number of members' responses in group g.

Axiom 2—Influencer.

If $\overline{N}_g < UR_i < 2 \times \overline{N}_g$ and $UNI < N_g$, then $user_i$ is an influencer. If the $user'_i$s responses (UR_i) are more than the average number of total responses in the group and lower than twice the average number of total responses in the group. Moreover, $user'_i$s speech which is NI and the count in matrix $T(US_i)$ which is less than 1 cannot more than the number of the member which is in the group. Influencers also have the power to influence others and have the potential to become opinion leaders. Hence, this study also offers a way to find influencers.

Axiom 3—Followers.

If $UR_i > 0$, then the user is a follower.

A follower needs to support or agree with someone, so the $user_i$ must have a positive count in matrix T. In other words, the user in row_i is the follower's leader.

Three social community patterns are classified: In the *criterion* pattern, the opinion leaders broadly influencing many followers usually obtain high social community support and posts professionally. The criterion social community, the most common pattern. In the argument pattern, the pattern's emergence is caused by the discussion space provided for users of social community platforms. Users can give specific advice to influence each other or influence other users. The bandwagon pattern arises when followers follow closely due to opinion leaders' personal charisma. In this pattern, followers usually do not care whether the content posted by opinion leaders is correct.

The following axioms are used to define three patterns.

Axiom 4—Criterion pattern.

According to *group_k* with opinion leaders in *set_ol*, (1) if the number of posts influencing followers who post to *set_ol* in *group_k* is more than half of the number ($\geq 1/2 \times SI\_f$), and (2) if the number of posts influencing opinions directed to followers in *group_k* is more than half of the number (SI_ol $\{HI,GI\} \geq 1/2 \times SI\_ol$), then *group_k* is a criterion pattern.

According to the criterion pattern, an enterprise can promote its products effectively, and the government can sharp public opinion in favor of a particular policy by utilizing the function of opinion leaders. Furthermore, finding opinion leaders and tracking them over the long term can prevent an explosion of potential issues. In the criterion pattern, the opinion leader is very professional.

If a government or enterprise wants to negotiate or cooperate with them, the contractor must also be professional.

Axiom 5—Argument pattern.

After grouping with the CI Algorithm, we find different groups. According to *group_k* with opinion leaders in *set_ol*, (1) if the number of posts influencing {HI,GI} followers who post to set_ol in group_k is more than half of the number (if *SI_f* {GI,HI} $\geq$ 1/2 $\times$ *SI_f*) and (2) if the number of posts influencing {HI, GI} opinions directed to followers in group_k is more than half of the number (*SI_ol* {HI,GI} $\geq$ 1/2 $\times$ *SI_ol*), then we recognize that group_k can be characterized as an argument pattern.

On the basis of the argument pattern, the character of the interaction between opinion leaders and followers is not significant; consequently, the cost of marketing is high and may even have little impact on promotion. Moreover, in the argument pattern, the viewpoints are diverse. Thus, it is desirable to provide a platform and sufficient information and domain knowledge for uses to engage in dialog with each other.

Axiom 6—Bandwagon pattern.

After grouping with the CI Algorithm, then we have different groups. According to *group_k* with opinion leaders in *set_ol*, (1) if the number of posts influencing {NI} followers who post to *set_ol* in *group_k* is more than half of the number (if *SI_f* {NI} $\geq$ 1/2 $\times$ *SI_f*) and (2) if the number of posts influencing {NI} opinions directed at followers in *group_k* is more than half of the number (*SI_ol* {NI} >1/2 $\times$ *SI_ol*), then we recognize that *group_k* can be characterized as a bandwagon pattern.

According to the bandwagon pattern, opinion leaders and followers are not professional in most cases. Enterprises and governments can utilize social media to promote their products and public policy effectively by enhancing the roles of these opinion leaders and followers. If the government or an enterprise wants to negotiate or cooperate with them, the contactors need not be professional but must be a decision-maker who can promise to provide resources.

3.3. Problem with Identification of Opinion Leaders and Followers

The problem of identifying opinion leaders and followers is formulated as follows:

Decompose a user-user interaction matrix into mutually separable submatrices (modules) with (1) the minimum number of non-empty high-value entries outside the block-diagonal matrix T, and (2) the maximum number of strongly desired entries (HI) and the minimum number of strongly undesired entries (NI) included in the submatrices of the block diagonal matrix I.

Subject to the following constraints:

Constraint C1: Empty groups of users are allowed, and

Constraint C2: The number of users in a group cannot exceed the upper bound Nu.

Constraint C3: Satisfy the following assumptions:

(1) Continuous posts are defined as one post.
(2) Users who respond negatively to posts cannot be regarded as followers.
(3) The content of the post and the level of social community support determine the influence of the user's post.
(4) Expert posts are prioritized as reasonable posts.
(5) If users have a low influence on each other, judge it as "NI".
(6) If users have a great influence on each other, judge it as "HI".

In matrix T, we count input post, responses and being responded to. In the matrix I, (1) input the highest influence power HI. Moreover, (2) input the lowest influence power NI. Combined with observation results, identify the opinion leader set and follower set.

3.4. Identification of Opinion Leader and Follower

In this study, the relationship between opinion leaders and followers is the mutual relationship between users. When a user satisfies the characteristics of opinion leaders, our study defines the user as an opinion leader. Followers will change their own ideas and attitudes according to opinion leaders' characteristics, including social status, accuracy of post contents and social community support level. The algorithm is described as follows:

Step I. Collect data from social media, such as users, posts, and response information.

Step II. Compute the counts of total posts of $user_i$ in matrix $[Tii]$

Step III. For all users, put the responses which $user_i$ gives to $user_j$ in matrix $[Tij]$ until there are no responses from $user_i$ to other users.

Step IV. If $colomn_i$ and row_i in matrix T are NULL then remove the meaningless $user_i$ by deleting row_i and $colomn_i$ in matrix T and I until there are no meaningless $user_i$

Step V. According to the data of social media, matrix T and social community support level, the social community support level marked with $user_i$ at $[Iii]$.

Step VI. For all users i and j, according to expert judgment, assign $user_i$ the influence power level $[Iij]$ of $user_j$.

Step VII. If $[T_{(i-1)(i-1)}] > [T_{ii}]$, exchange the columns of $user_i$ and $user_{i-1}$. until there is no $[T_{ii}] < [T_{(i-1)(i-1)}]$ in matrix T.

Step VIII. The CI algorithm is applied to group users.

Step IX. If $[Iij]$ is not NULL, then check whether the $user_i$ and $user_j$ is in the same group or not; if they are not in the same group, then put them in the same group matrix until all users in the matrix I have been checked.

Step X. In each group, sum up all positive responses to N_{rsp} and compute the average $\overline{N}_{rsp}$.

Step XI. Count the responses of $user_i$ by $\sum_{j=1}^{n}\left[T_{ij}\right]$ and posts of $user_i$ by $\sum_{i=1}^{n}\left[T_{ij}\right]$. Additionally, count the influence of each $user_i$, If $user_i$ satisfies Axiom 1, then $user_i$ is identified as an opinion leader. If $user_i$ satisfies Axiom 2, then $user_i$ is identified as an influencer. If $user_i$ satisfies Axiom 3, then $user_i$ is identified as a follower. Continue until all users in matrix T have been checked.

Step XII. Check each group matrix. Count all opinion leaders' $[I_{ij}]$ and followers' $[I_{ij}]$. Recognize the pattern based on Axioms 4–6.

Step XIII. When there are positive responses or influence between users, we classify these users in the same group.

4. Case Study

The ABC network platform is taken as an example to describe the application of our research in practice. The ABC network platform is a discussion platform created by the government to promote community communication. This platform was created as part of a public policy proposal to improve policy communication and make policy public.

This case study is taken from the National Energy Conference organized by the Energy Bureau of Taiwan's Ministry of Economic Affairs. However, there are still many disagreements when it comes to choosing opinions due to value divergence. To discuss and clarify issues with the public, the proposition, "Where does future electric power come from?" is open on the policy consultant forum (People Talk), with three sub-issue forums including, 'environment low carbon sustainable development', 'stable supply and open source' and 'reduce expenditures effectively'. In particular, 'stable supply and open source' is the focus of this case study.

The proposed solution approach is applied in this case.

Step I. Collect materials: Judging by the forum (posts, fan pages) on social media about green energy and low carbon, we collected materials, including text and response information. This study collected materials from users' discussion contents related to the "stable supply" issue on the ABC network platform between May 2019 and the end of 2019. The data collection is implemented with the Python-Jieba crawler program, which is particularly suitable for Chinese text analysis automatically. The collected materials are listed below: Post users: 36; total posts: 205 (total posts have been deducted from the number of administrator responses and consecutive posts); effective responses: 61; effective count of being responded to 47.

Step II. Input matrix elements: Input elements in matrix *T*. The elements include general posts, counts of responses and being responded to as judged by experts.

Step III. Remove meaningless users: Remove users whose posts are never responded to.

Step IV. Tag the user's category: Input social community support

Step V. Influence analysis: We analyze a user's influence by comparing the levels of influence power according to three characteristics and input the influence into the matrix *I*.

Step VI. If the count of users' posts, responses, and influence power, reaches a certain level of relevance, then move forward. If the users have greater counts of responses or respondents, list them in front. (Figure 1).

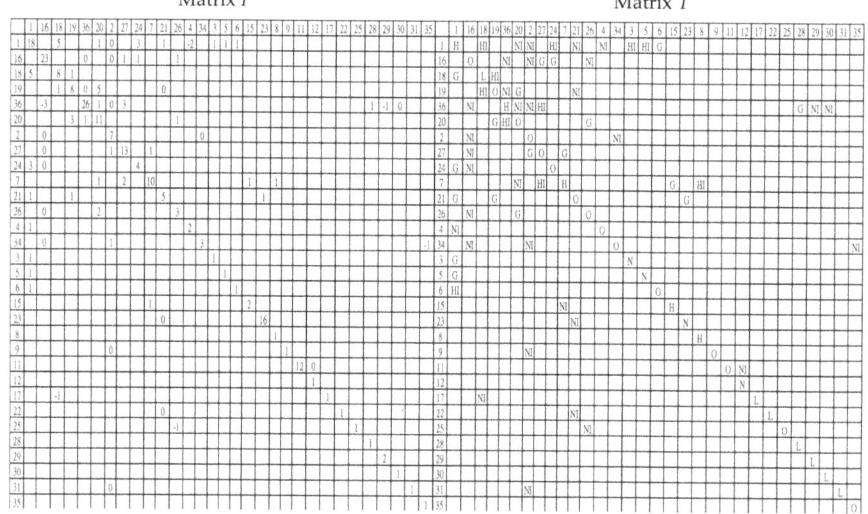

Figure 1. Matrix element conversion. Note1: the criterion of influence power: NI = no influence; GI = general influence; HI = high influence. Note2: The level of social community support: H: high social community support; L: low social community support; O: missing social community support.

Step VII. According to matrix *I* and *T* based on Equation (1), group the users by their relationships. Group A {1, 2, 3, 4, 5, 6, 18, 20, 21, 24}; Group B {2, 16, 24, 26, 27, 34, 36}; Group C {17, 18, 19}; Group D {19, 20, 21, 36}; Group E {2, 27, 28, 29, 30, 36}; Group F {7, 20, 26}; Group G {2, 9, 27, 31, 34}; Group H {7, 27}; Group I {7, 8, 15}; Group J {21, 22, 23}; Group K {25, 26}; Group L {34, 35}.

Step VIII. Identification: Determine the interaction between users, followers, and opinion leaders, according to the definition of each group of opinion leaders.

In Group A, opinion leader 1 is recognized by Axiom 1, and the influencers are Users 3 and 5. This group is identified as an argument group based on Axiom 5.

In Group B, User 16's post contents are usually meaningless. According to Axiom 1, User 16 is not an opinion leader. Moreover, the group does not belong to any pattern.

Group C is an argument pattern. Users 18 and 19 are influencers. In this pattern, no opinion leaders and followers are identified. The influencers influence each other without focusing on any particular key person.

In group D, the response of User 19 has a great influence on Users 20 and 21. However, the influence of the posts responded to by Users 20 and 21 is not great. There is a discussion relationship between Users 19 and 20, so it is a social pattern.

In Group E, the post contents of User 36 are valuable. However, other users' responses are not good. Thus, User 36 is an opposing opinion leader.

In Group F, according to Axiom 6, User 26 is an influencer, and the group represents an argument pattern. User 20 is the follower of User 26.

In Group G, Users 27 and 34 are the influencers, and it is a bandwagon pattern. Moreover, User 2 is the follower.

In Group H, Users 27 and 7 have a discussion relationship, and both of them are influencers.

In Group I, the social community support level of the group is high. They should be opinion leaders, in theory. However, in this case, study, since they do not play the role of opinion leaders, they cannot be recognized as opinion leaders. User 7 is an opinion leader, and this is an argument pattern.

In Group J, there are no opinion leaders or followers. User 21 is an influencer.

In Group K, User 26 is opposed to the opinions of User 25. There are no opinion leaders or followers in this group.

Group L: In this group, there are no opinion leaders or followers.

According to the summary of group analysis, Users 1 and 36 are obviously opinion leaders. The five groups A, C, D, F, and H can all be characterized as argument patterns, which shows that in the forum, most post contents influence other users through discussion.

The case was also analyzed with a traditional network approach, i.e., the Ward method, named after its creator, focuses on the allocation of profiles to groups equally. Ward [50] pointed out that grouping in this manner makes it easier to consider and understand relations in large collections. The principle of this method is to minimize heterogeneity, and the important goal is to find the greatest similarity. The comparison between the proposed approach and Ward's approach is shown in Table 3. The results show that Users 1, 7, 36 are identified as opinion leaders. However, User 19 has not been identified through the traditional method due to the threshold.

Table 3. Analysis results.

		The Proposed Approach		Ward's Approach
Opinion leader		1,7,19,36		1,7,36
Influencer		18,24,20,21,22,27		N/A
Leader/lollowers	1	18,24,21,3,5,6,20	1	18,24,21,3,5,6,20
	7	8,15,20,27	7	8,15,20,27
	19	8,15,20	36	20,27,28
	36	20,27,28		
Pattern	Criterion	D, E		
	argument	A, C, F, H, I		N/A
	bandwagon	G		

The identification of opinion leaders by Ward's method only identifies opinion leaders who participate in the whole conversation. However, in the proposed approach, this study uses two identification methods: the whole conversation and group conversation. The latter can clarify which user is the group's opinion leader. In addition, the proposed approach can discover different patterns. Although the traditional network approach of Ward's is considered to be the best one among the hierarchical clustering methods [51–53], it cannot identify these patterns.

Through the perception of social community patterns among users, this study successfully distinguished opinion leader and defined social patterns in the complex social communities, which contains highly controversial users and many of them are anonymous, where few persons are involved in the discussion and users' support level could not be obtained because users disagree with each other.

After identifying opinion leaders and followers, in a criterion pattern, opinion leaders have a higher degree of professionalism than followers. In that case, if green energy and low-carbon related policies are to be disseminated through opinion leaders, it is necessary to send personnel with a certain degree of professionalism. After contacting and negotiating with them, you must first obtain the approval of the opinion leaders before you can persuade them to influence followers through their platforms or social media. It is expected that they will achieve rapid dissemination, higher dissemination effect, and avoid costly but ineffective dissemination. In addition, a follower may also become another opinion leader, generating multiple diffusion of innovations.

Second, in the argument pattern, due to the comparably equal status between opinion leaders and followers, issues are quite diverse, and it is not easy to focus on specific issues. If opinion leaders wanted to disseminate relevant policies on energy conservation and low carbon to influence followers, the dissemination effect would be poor. Therefore, in order to make the issue of green energy and low-carbon attract more attention, opinion leaders can package the issue into lifestyles and features, thereby achieving a higher diffusion effect (Diffusion of innovations) on followers.

In addition, in the bandwagon pattern, because opinion leaders and followers are less professional, they are more vulnerable to each other. In order to disseminate green energy and low-carbon policies, policies can be packaged as simple, interesting or lifestyle issues, while social media or platforms are often used by opinion leaders or followers to achieve better diffusion of innovations.

5. Discussion

In this study, three interactive patterns and their characteristics are identified, which can help how to find opinion leaders more effectively and grasp the characteristics of opinion leaders and followers when want to spread (Diffusion of innovation) new policies or marketing new products. Opinion leaders and followers both have different levels of knowledge, social community support, and influence power. Therefore, this study summarizes the interactions on social media into three patterns, and the characteristics of three patterns have also been explored. Furthermore, based on the characteristics of users in these patterns, it can be used to provide opinion leaders with specific and clear topics/issues to influence their followers, thereby obtaining effective dissemination or commercial marketing purposes in the green energy domain.

In addition, the results of this study can also be applied to the political dissemination of democratic elections or the shaping of the opinion climate, which can more efficiently lead the electoral issues and win elections. In other words, the issues or political opinions that candidates are trying to market can be differentiated based on different communication modes so that the information can be segmented, and the impact of effective agenda-setting goals can be achieved. This study not only has the possibility of expanding and deeper research, but it is also the relative value of this research.

Most previous studies used different algorithms or improved algorithms to identify opinion leaders [28–31,33,37]. In other words, most of the above-mentioned studies only used various algorithms to identify opinion leaders or followers, and consequently, apply them to political communication and commercial marketing related fields. There has recently been an exploration of the psychological motivation of actively acting as opinion leaders to understand which users are active communicators or passive recipients of social issues. However, the related study on the interaction patterns between the opinion leaders and followers and their characteristics have not been explored.

In addition, the opinion dynamics of current popular research are interested. The classic model of opinion dynamics is derived from the research of DeGroot's and Friedkin–Johnsen's models of opinion dynamics, which aims at the integration and consistency of opinions in social networks, carried out very enlightening modeling and exploration [54]. DeGroot's model describes the process of reaching consensus in social networks, while Friedkin–Johnsen's model further introduces the degree of "stubborn individual" to explain the phenomenon of inconsistent opinions in social networks. The models clearly depict the dynamic process of opinion integration and consistency, as well as the obstacles caused by "stubborn individual" factors to the process of opinion integration [55,56]. However, the two models are very instructive to explore how individuals (or Internet users) should be controlled if they are affected by certain characteristics or stubbornness in the process of opinion integration. Our study more specifically explores how opinion leaders and followers can find out the characteristics of users and the patterns of interaction between them in the process of consensus, which can be applied to precision marketing in actual operation, and even give opinion leaders with different characteristics use differentiated topic content to increase the influence of consensus. Since the research of DeGroot's and Friedkin–Johnsen's models of opinion dynamics are in conceptual level only, some difficulties in practical application are challenged [57].

This study extends the idea of [39] to the domain of green energy and low carbons, where roughly qualitative characteristics of opinion leaders and matrix of interaction between users are considered. To make this study more solid and applicable, the theories of "two-step flow", "bandwagon effect", "agenda-setting" and "innovation diffusion theory" from the theoretical perspective of communication and the axioms are used to validate the results. In addition, this study does not focus on a single discipline only but a cross-disciplinary study of the fields of green energy and low carbons, intelligent systems, and communication to provide numerous management implications discussed in this section. The core novelty and contribution is shown in that the solid theoretical part makes this study applicable to other social media and industry sectors.

6. Conclusions

Nowadays, networks are the most important media among broad masses, and almost everyone is closely related to networks, which results in many social issues as virtual networks are reflected in the real world. Opinion leaders play a very important role in spreading media for many issues. Previous research used some social measurement methods based on interview self-reports or questionnaire surveys [23,24,32,33] or used quantitative clustering techniques to identify opinion leaders [30,40–43]. In fact, few studies have investigated online posts to determine the social patterns of opinion leaders and interactions between opinion leaders and followers. Therefore, it is valuable that this study proposes a method to analyze the characteristics of social media posts, posters and their interaction relationships.

Furthermore, this study identifies opinion leaders on the issue of green power in social communities based on social community support level and influence power level. As a result of users cannot express positive and negative opinions on the issue; it is not enough to consider only the user's posts. For example, when a user has a large number of posts, if the user cannot get support from others, he/she cannot be classified as an opinion leader.

Using the same model and the solution approach, the results of this study can be extended from the green energy low carbon issue to other social issues, e.g., the domain of marketing to make better CRM. It also provides an efficient and operable application model for online marketing or public issue communication in practical applications, which can more easily identify opinion leaders and followers. Furthermore, this study not only examines from the perspective of intelligent systems such as different algorithms but provides a new perspective to deal with the effective identification of opinion leaders and followers, at the same time, promotes the "theory of two-step flow" [5] to add another research perspective in the academic field.

The level of community support and influence proposed in this study uses a relational matrix to analyze the relationship and the community pattern between users in this study. If it is applied to analyze social media with higher data volume and discussion volume, it should consider the computation speed, but which is usually not an issue in the current IT world. In addition, this study uses static data for analysis. This model can also be added or removed from dynamic data and evolution modeling for analysis in the future, and it is expected that it can be identified as more timely and faster. For example, the advantages of DeGroot and Friedkin–Johnsen models are taken into consideration for further study

However, this research's reproducibility to other industry sectors is interested and requires further investigation. In addition, this study focuses on a single case study of a country's green energy and low-carbon policy in Taiwan. If the results of this study are used to infer whether there will be differences in other countries with different levels of development, knowledge and education, it is worthwhile to explore further. In any case, despite the above negotiable points, it does not detract from the valuable results obtained in this study.

Author Contributions: Conceptualization, W.-Y.L.; methodology development, C.-C.H.; coding, P.-A.C.; data preparation and analysis, Y.-C.C.; writing—review and editing, W.-Y.L. and C.-C.H.; data collection and annotation, P.-A.C. and Y.-C.C. All authors have read and agreed to the published version of the manuscript.

Funding: This research was funded by the National Science Foundation—Ministry of Science and Technology of Taiwan, grant number MOST-104-3113-F-260-001 and MOST 107-2410-H-260 -014 -MY3.

Conflicts of Interest: The authors declare no conflict of interest.

References

1. Al-Hasan, A.; Yim, D.; Lucas, H.C. A tale of two movements: Egypt during the Arab spring and occupy wall street. *IEEE Trans. Eng. Manag.* **2018**, *66*, 84–97. [CrossRef]
2. Kietzmann, J.H.; Hermkens, K.; McCarthy, I.P.; Silvestre, B.S. Social media? Get serious! Understanding the functional building blocks of social media. *Bus. Horiz.* **2011**, *54*, 241–251. [CrossRef]
3. Ang, L. Community relationship management and social media. *J. Database Mark. Cust. Strategy Manag.* **2011**, *18*, 31–38. [CrossRef]
4. Ramonet, I.; Moreno, G.R. *Fundación para la Investigación y la Cultura*; El Quinto Poder: Santiago, Chile, 2004.
5. Katz, E.; Lazarsfeld, P.F. *Personal Influence, The Part Played by People in the Flow of Mass Communications*; Transaction Publishers: Piscataway, NJ, USA, 1966.
6. Rosen, E. *The Anatomy of Buzz Revisited: Real-Life Lessons in Word-of-Mouth Marketing*; Doubleday: New York, NY, USA, 2009.
7. Elghali, S.B.; Benbouzid, M.; Charpentier, J.F. Marine tidal current electric power generation technology: State of the art and current status. In Proceedings of the 2007 IEEE International Electric Machines & Drives Conference, Antalya, Turkey, 3–5 May 2007; pp. 1407–1412.
8. Shun, T.-Y. A Study of Opinion Leaders and Communication Features of Blogs. Master's Thesis, National Taipei University of Technology, Taipei, Taiwan, 2006.
9. Tachizawa, E.M.M.; Thomsen, C.G.; Montes-Sancho, M.J. Green supply management strategies in Spanish firms. *IEEE Trans. Eng. Manag.* **2012**, *59*, 741–752. [CrossRef]
10. Corbett, J.; Wardle, K.; Chen, C. Toward a sustainable modern electricity grid: The effects of smart metering and program investments on demand-side management performance in the US electricity sector 2009–2012. *IEEE Trans. Eng. Manag.* **2018**, *65*, 252–263. [CrossRef]
11. Maurovich-Horvat, L.; De Reyck, B.; Rocha, P.; Siddiqui, A.S. Optimal selection of distributed energy resources under uncertainty and risk aversion. *IEEE Trans. Eng. Manag.* **2016**, *63*, 462–474. [CrossRef]
12. Akhegaonkar, S.; Nouveliere, L.; Glaser, S.; Holzmann, F. Smart and green ACC: Energy and safety optimization strategies for EVs. *IEEE Trans. Syst. Man Cybern. Syst.* **2016**, *48*, 142–153. [CrossRef]
13. Höök, M.; Tang, X. Depletion of fossil fuels and anthropogenic climate change—A review. *Energy Policy* **2013**, *52*, 797–809. [CrossRef]
14. Yildirim, M.B.; Mouzon, G. Single-machine sustainable production planning to minimize total energy consumption and total completion time using a multiple objective genetic algorithm. *IEEE Trans. Eng. Manag.* **2011**, *59*, 585–597. [CrossRef]
15. Midilli, A.; Dincer, I.; Ay, M. Green energy strategies for sustainable development. *Energy Policy* **2006**, *34*, 3623–3633. [CrossRef]
16. Li, J.; Ying, Y.; Lou, X.; Fan, J.; Chen, Y.; Bi, D. Integrated energy system optimization based on standardized matrix modeling method. *Appl. Sci.* **2018**, *8*, 2372. [CrossRef]
17. Xu, G.; Wu, S.; Tan, Y. Island partition of distribution system with distributed generators considering protection of vulnerable nodes. *Appl. Sci.* **2017**, *7*, 1057. [CrossRef]
18. Gitau, J.K.; Mutune, J.; Sundberg, C.; Mendum, R.; Njenga, M. Implications on Livelihoods and the Environment of Uptake of Gasifier Cook Stoves among Kenya's Rural Households. *Appl. Sci.* **2019**, *9*, 1205. [CrossRef]
19. Huang, F.L. The Trend and Key Successful Factors for Taiwan Green Industry. Master's Thesis, National Taipei University of Technology, Taipei, Taiwan, 2015.

20. Liu, P.W. Exploring Internet Policy Opinion in the Era of Big Data: A Case Study of Free Economic Pilot Zones in Taiwan. Master's Thesis, National Chengchi University, Taipei, Taiwan, 2016.

21. Grissa, K. What makes opinion leaders share brand content on professional networking sites (e.g., LinkedIn, Viadeo, Xing, SkilledAfricans …). In Proceedings of the 2016 International Conference on Digital Economy (ICDEc), Carthage, Tunisia, 28–30 April 2016; pp. 8–15.

22. Hartigan, J.A.; Wong, M.A. Algorithm AS 136: A k-means clustering algorithm. *J. R. Stat. Soc. Ser. C (Appl. Stat.)* **1979**, *28*, 100–108. [CrossRef]

23. Li, F.; Du, T.C. Who is talking? An ontology-based opinion leader identification framework for word-of-mouth marketing in online social blogs. *Decis. Support Syst.* **2011**, *51*, 190–197. [CrossRef]

24. Cho, Y.; Hwang, J.; Lee, D. Identification of effective opinion leaders in the diffusion of technological innovation: A social network approach. *Technol. Forecast. Soc. Chang.* **2012**, *79*, 97–106. [CrossRef]

25. Li, J.; Xing, G.; Wang, Y.; Ren, Y. Training opinion leaders in microblog: A game theory approach. In Proceedings of the 2012 Second International Conference on Cloud and Green Computing, Xiangtan, China, 1–3 November 2012; pp. 754–759.

26. Ali, M.; Baqir, A.; Psaila, G.; Malik, S. Towards the Discovery of Influencers to Follow in Micro-Blogs (Twitter) by Detecting Topics in Posted Messages (Tweets). *Appl. Sci.* **2020**, *10*, 5715. [CrossRef]

27. Burt, R.S. The social capital of opinion leaders. *Ann. Am. Acad. Political Soc. Sci.* **1999**, *566*, 37–54. [CrossRef]

28. Duan, J.; Zeng, J.; Luo, B. Identification of opinion leaders based on user clustering and sentiment analysis. In Proceedings of the 2014 IEEE/WIC/ACM International Joint Conferences on Web Intelligence (WI) and Intelligent Agent Technologies (IAT), Warsaw, Poland, 11–14 August 2014; pp. 377–383.

29. Weng, J.; Lim, E.-P.; Jiang, J.; He, Q. Twitterrank: Finding topic-sensitive influential twitterers. In Proceedings of the Third ACM International Conference on Web Search and Data Mining, New York, NY, USA, 3–6 February 2010; pp. 261–270.

30. Ziyi, L.; Jing, C.F.S.; Donghong, S.; Yongfeng, H. Research on methods to identify the opinion leaders in Internet community. In Proceedings of the 2013 IEEE 4th International Conference on Software Engineering and Service Science, Beijing, China, 23–25 May 2013; pp. 934–937.

31. Song, X.; Chi, Y.; Hino, K.; Tseng, B. Identifying opinion leaders in the blogosphere. In Proceedings of the Sixteenth ACM Conference on Conference on Information and Knowledge Management, Lisbon, Portugal, 6–10 November 2007; pp. 971–974.

32. Kim, B.; Kang, S.; Lee, S. A Weighted PageRank-Based Bug Report Summarization Method Using Bug Report Relationships. *Appl. Sci.* **2019**, *9*, 5427. [CrossRef]

33. Jing, L.; Lizhen, X. Identification of microblog opinion leader based on user feature and interaction network. In Proceedings of the 2014 11th Web Information System and Application Conference, Tianjin, China, 12–14 September 2014; pp. 125–130.

34. Tørning, K.; Jaffari, Z.; Vatrapu, R. Current challenges in social media management. In Proceedings of the 2015 International Conference on Social Media & Society, Toronto, ON, Canada, 27–29 July 2015; pp. 1–6.

35. Ben-Hur, A.; Horn, D.; Siegelmann, H.T.; Vapnik, V. Support vector clustering. *J. Mach. Learn. Res.* **2001**, *2*, 125–137. [CrossRef]

36. Venkatraman, M.P. Opinion leaders, adopters, and communicative adopters: A role analysis. *Psychol. Mark.* **1989**, *6*, 51–68. [CrossRef]

37. Kaufman, L.; Rousseeuw, P. *Finding Groups in Data; an Introduction to Cluster Analysis*; 0471878766; John Wiley & Sons: Hoboken, NJ, USA, 1990.

38. Bezdek, J.C.; Ehrlich, R.; Full, W. FCM: The fuzzy c-means clustering algorithm. *Comput. Geosci.* **1984**, *10*, 191–203. [CrossRef]

39. Huang, C.-C.; Lien, L.-C.; Chen, P.-A.; Tseng, T.-L.; Lin, S.-H. Identification of Opinion Leaders and Followers in Social Media. In Proceedings of the DATA, Madrid, Spain, 24–26 July 2017; pp. 180–185.

40. Ma, N.; Liu, Y. SuperedgeRank algorithm and its application in identifying opinion leader of online public opinion supernetwork. *Expert Syst. Appl.* **2014**, *41*, 1357–1368. [CrossRef]

41. Jiang, L.; Ge, B.; Xiao, W.; Gao, M. BBS opinion leader mining based on an improved PageRank algorithm using MapReduce. In Proceedings of the 2013 Chinese Automation Congress, Changsha, China, 7–8 November 2013; pp. 392–396.

42. Cheng, F.; Yan, C.; Huang, Y.; Zhou, L. Algorithm of identifying opinion leaders in BBS. In Proceedings of the 2012 IEEE 2nd International Conference on Cloud Computing and Intelligence Systems, Hangzhou, China, 30 October–1 November 2012; pp. 1149–1152.

43. Deng, X.; Li, Y.; Lin, S. Parallel Micro Blog Crawler Construction for Effective Opinion Leader Approximation. *AASRI Procedia* **2013**, *5*, 170–176. [CrossRef]

44. Chu, K.-H.; Majmundar, A.; Allem, J.-P.; Soto, D.W.; Cruz, T.B.; Unger, J.B. Tobacco use behaviors, attitudes, and demographic characteristics of tobacco opinion leaders and their followers: Twitter analysis. *J. Med. Internet Res.* **2019**, *21*, e12676. [CrossRef] [PubMed]

45. Lin, H.-C.; Bruning, P.F.; Swarna, H. Using online opinion leaders to promote the hedonic and utilitarian value of products and services. *Bus. Horiz.* **2018**, *61*, 431–442. [CrossRef]

46. Zhao, Y.; Kou, G.; Peng, Y.; Chen, Y. Understanding influence power of opinion leaders in e-commerce networks: An opinion dynamics theory perspective. *Inf. Sci.* **2018**, *426*, 131–147. [CrossRef]

47. Winter, S.; Neubaum, G. Examining characteristics of opinion leaders in social media: A motivational approach. *Soc. Media Soc.* **2016**, *2*, 2056305116665858. [CrossRef]

48. Paul, D.; Volinsky, C. Method and Apparatus to Identify Influencers. US Patent 9,946,975, 17 April 2018.

49. Khan, N.S.; Ata, M.; Rajput, Q. Identification of opinion leaders in social network. In Proceedings of the 2015 International Conference on Information and Communication Technologies (ICICT), Karachi, Pakistan, 12–13 December 2015; pp. 1–6.

50. Ward Jr, J.H. Hierarchical grouping to optimize an objective function. *J. Am. Stat. Assoc.* **1963**, *58*, 236–244. [CrossRef]

51. Eszergár-Kiss, D.; Caesar, B. Definition of user groups applying Ward's method. *Transp. Res. Procedia* **2017**, *22*, 25–34. [CrossRef]

52. Majerova, I.; Nevima, J. The measurement of human development using the Ward method of cluster analysis. *J. Int. Stud.* **2017**, *10*, 239–257. [CrossRef] [PubMed]

53. Hale, R.L.; Dougherty, D. Differences between ward's and UPGMA methods of cluster analysis: Implications for school psychology. *J. Sch. Psychol.* **1988**, *26*, 121–131. [CrossRef]

54. DeGroot, M.H. Reaching a consensus. *J. Am. Stat. Assoc.* **1974**, *69*, 118–121. [CrossRef]

55. Rey, S.I.; Reyes, P.; Silva, A. Evolution of social power for opinion dynamics networks. In Proceedings of the 2017 55th Annual Allerton Conference on Communication, Control, and Computing (Allerton), Monticello, IL, USA, 3–6 October 2017; pp. 716–723.

56. Ravazzi, C.; Hojjatinia, S.; Lagoa, C.M.; Dabbene, F. Randomized opinion dynamics over networks: Influence estimation from partial observations. In Proceedings of the 2018 IEEE Conference on Decision and Control (CDC), Miami Beach, FL, USA, 17–19 December 2018; pp. 2452–2457.

57. Zhou, Q.; Wu, Z.; Altalhi, A.H.; Herrera, F. A two-step communication opinion dynamics model with self-persistence and influence index for social networks based on the degroot model. *Inf. Sci.* **2020**, *519*, 363–381. [CrossRef]

Publisher's Note: MDPI stays neutral with regard to jurisdictional claims in published maps and institutional affiliations.

MDPI

St. Alban-Anlage 66

4052 Basel

Switzerland

Tel. +41 61 683 77 34

Fax +41 61 302 89 18

www.mdpi.com

Applied Sciences Editorial Office

E-mail: applsci@mdpi.com

www.mdpi.com/journal/applsci